PLANTS, PEOPLE, AND PALEOECOLOGY

Biotic Communities and Aboriginal Plant Usage in Illinois

by

FRANCES B. KING

ILLINOIS STATE MUSEUM
Scientific Papers, Vol. XX

Illinois State Museum
Springfield, Illinois
1984

STATE OF ILLINOIS
Department of Energy and Natural Resources

ILLINOIS STATE MUSEUM
R. Bruce McMillan, Ph.D., Museum Director

BOARD OF THE
ILLINOIS STATE MUSEUM

Michael G. Schneiderman, Chairman
Chicago

Don Etchison, Director
Department of Energy and
Natural Resources

James Ballowe, Ph.D.
Peoria

Rosalie Clark
Chicago

Donald F. Hoffmeister, Ph.D.
Urbana

George M. Irwin
Quincy

Mary Ann MacLean
Libertyville

Jane O'Connor
Chicago

Robert H. Waddell
Illiopolis

Sol Weiner
Evanston

1984
ISSN 0445-3395
ISBN 0-89792-100-3
Printed by Authority of the State of Illinois

CONTENTS

FIGURES

TABLES

PREFACE

Plants have the unique ability to absorb water, minerals, and sunlight and convert them into living matter which, in turn, provides the basic food source for other types of organisms, including man. In addition to forming the bulk of our diets, plants are of utmost importance to the comfort and well-being of persons of every level of sophistication and economic status. They yield wood for fuel, houses, and tools; fibers for fabrics and paper; dyes, oils, and chemical compounds used in medicines, soaps, and pesticides; and a multitude of other applications. Plants are the primary producers in our world; without them life would not exist as we know it.

The objectives of this study are severalfold. Prior to this, there has been no publication listing the edible plants or discussing those plants with potential aboriginal use in Illinois. Although some local studies exist for portions of the state, it has generally been necessary to extrapolate data from studies in adjacent regions. This study, therefore, is meant as a ready reference on the distribution and use of Midwestern wild or naturalized plants (those that were introduced in one manner or another and then became established in the wild) for foods, medicines, beverages, tobacco substitutes, dyes or technological purposes. While written primarily for archaeologists, the information presented here should be of value to persons interested in natural dyeing, edible wild plants, herbal teas, or "natural" medicines.

Subsistence models are being built with increasing frequency by archaeologists in an attempt to maximize their understanding of prehistoric human societies; it is important, therefore, that the accuracy of the ecological data base supporting such models be as great as possible. At the same time, there is a large body of published literature on the ecology of Illinois that archaeologists seldom utilize. The discussions of the natural divisions and natural communities of Illinois, as well as numerous references, are included to increase familiarity with the ecological data base that is available. The discussions of presettlement vegetation, past climates, and the environmental factors determining the structure and composition of modern plant communities may aid in the reconstruction and interpretation of the environment faced by the prehistoric occupants of Illinois.

The first section of this publication deals with the environmental factors controlling the plant communities of Illinois, particularly geology and climate. An understanding of these factors and their relationship to vegetation is an essential prerequisite to the reconstruction and interpretation of past vegetational communities and human subsistence based on the exploitation of those communities.

The second section lists the plants reported to have been used by the Historic Indians of the region. More recent uses by the Euro-American inhabitants are also given. In all cases, references to the original sources are included as is information on distribution, abundance, and habitats that should aid in the reconstruction of resource availability and the interpretation of subsistence patterns of the long-vanished occupants of specific archaeological sites.

The final section gives a brief description of the natural communities of Illinois, including the dominant and characteristic plants of each and a bit about their distribution. In addition, the potential food plants are listed for 63 common community types.

ACKNOWLEDGMENTS

I wish to thank the many people who contributed to this study. Mary Ann Graham spent many tedious hours . Drs. J. E. King, R. B. McMillan, B. W. Styles, and M. D. Wiant of the Illinois State Museum and P. J. Watson of Washington University, St. Louis, contributed ideas through discussion or by comments on the manuscript. Doris Epstein of Washington University contributed both suggestions and permission to cite her Master's thesis on plants used for smoking materials. Dr. A. C. Koelling, Illinois State Museum, undertook the onerous task of reviewing the scientific nomenclature. Orvetta Robinson edited the manuscript and saw it through publication as one of the first attempted at this institution from word processor files. The cover was designed by Julianne Snider. In addition to these people, I thank all of the archaeologists who have asked me to read, write, or edit discussions of environment and of aboriginal plant usage; for it is they who made the need for this study apparent.

THE ENVIRONMENTAL FRAMEWORK FOR UNDERSTANDING ABORIGINAL SUBSISTENCE IN ILLINOIS

Illinois forms a portion of the Great Central Plain of North America. The maximum length of the state is 380 miles north-south and the maximum width is more than 200 miles. Illinois is situated between 37 degrees and 42 degrees north latitude and 87 degrees and 91 degrees west longitude, covering approximately 57,926 square miles. The total relief of Illinois is about 950 feet with the average elevation about 600 feet above sea level. Although most of the state is relatively level, there is a well-developed drainage system of approximately 275 streams that ultimately flow into the Mississippi, Wabash, and Ohio rivers (G. Jones 1950: 2-3).

The flora (as well as the fauna) of Illinois is similar to that of the adjacent states (Fig. 1), reflecting climatic and geologic similarities. Several major vegetation regions converge in Illinois. These include prairie and beech-maple, maple-basswood, oak-hickory, western-mesophytic, and southern forest types (Braun 1950).

At the time of Euro-American settlement, the northern two-thirds of the state was covered by broad prairies with tongues of forest encroaching along the larger streams. The great stretches of prairie, comprised of rich growths of tall grasses interspersed with hundreds of species of herbaceous plants, were one of the most remarkable features of the early Illinois landscape. On the more poorly drained, formerly glaciated surfaces, prairies were broken by marshes and ponds. Certain areas of central and northern Illinois have sandy soils that continue to support characteristic "sand prairie" species, many of which are more common on the drier Great Plains to the west. Forest occurred chiefly in the southern part of the state, on the floodplains of the larger rivers, and in the Ozark Hills. These forests were, and are, comprised almost entirely of hardwoods, of which oak, hickory, maple, elm, and ash are most common (G. Jones 1950: 3).

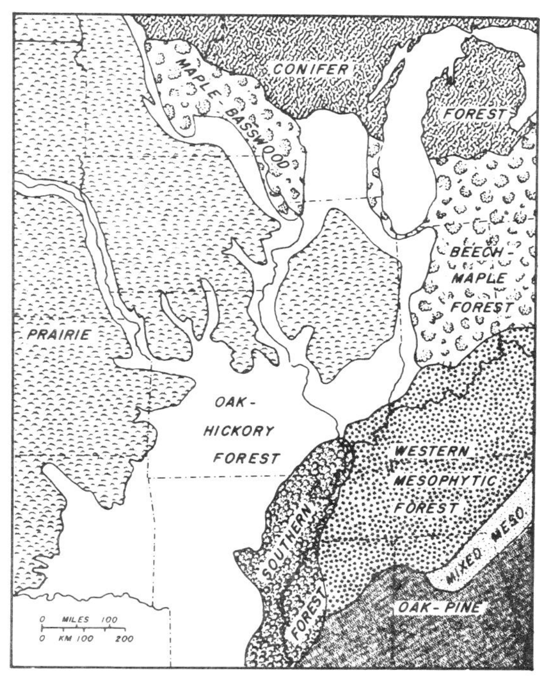

1. Forest Regions of the Midwest (after Braun 1950).

GEOLOGY

The geology of Illinois is the primary determinant of both topography and soils, two of the principal factors influencing the distribution and composition of plant communities. The structure of bedrock in the state forms a spoon-shaped depression, called the Illinois Basin, with a maximum depth occurring near the mouth of the Wabash River in southeastern Illinois. Formation of the Illinois Basin resulted from repeated tectonic movements apparently beginning during the Cambrian Period of the Paleozoic Era (about 570 million years ago) and continuing until the late Pennsylvanian (about 280 million years ago). During this time, the basin was open to seas advancing from the south (Willman *et al.* 1975:23).

The uppermost bedrock strata over much of Illinois are predominantly marine sediments (limestones) deposited during the

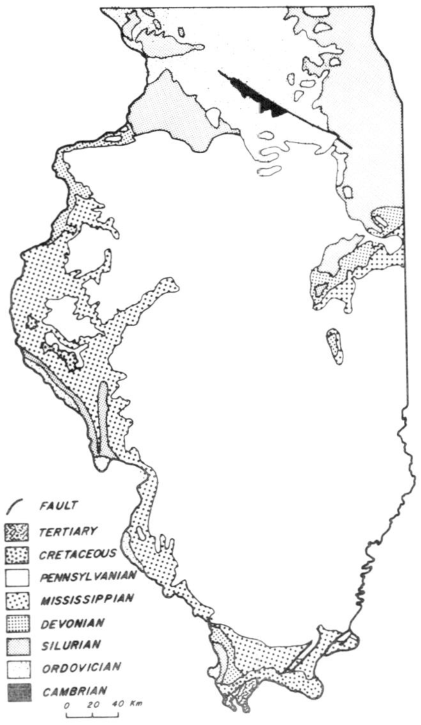

2 Geologic map of Illinois (after Willman and Frye, 1970.

Paleozoic (Fig. 2). The majority of this bedrock represents the most recent period of the Paleozoic, the Pennsylvanian. By this time, plants were abundant on the land surface, and the Pennsylvanian Period in Illinois was a time of warm, shallow seas and fresh-water swamps. Forests existing on broad deltaic plains were periodically submerged and ultimately buried due to rises in sea level and/or subsidence of the underlying basin. Eventually, the buried organic debris from these previous forests formed the coal beds. There are approximately 75 coal lenses, some widespread, representing the Pennsylvanian Period in Illinois (Willman *et al.* 1975:10,33). Virtually all of the areas underlain by Pennsylvanian bedrock have coal deposits at some depth.

The Mesozoic Era which followed the Paleozoic is represented only by relatively small areas of Cretaceous Period materials in southern and western Illinois (primarily marine deltaic or nearshore deposits). Cenozoic Era deposits, including those accumulating at the present, are the surficial materials over almost the entire state (Willman *et al.* 1975:19).

TOPOGRAPHY

Topography is a key factor in the distribution of biotic habitats. In the glaciated regions, the moraines and dissected stream valleys provide situations in which trees can successfully compete against the thick prairie sod. Broken topography provides protection not only from sun and wind desiccation but also from the prairie fires that frequently swept across the landscape during presettlement times. Rugged landscapes also result in an increasingly large number of microenvironments. These range from creeks and shaded ravines to bedrock outcrops and steep, exposed slopes. It is not uncommon to find both moist- and dry-adapted species growing within a few feet of one another when the topography is broken and the landscape varied.

Except for small portions of northwestern, west-central and southern Illinois, the state is covered by glacially deposited material, including ground and end moraines, glacial outwash and glacial lake deposits. Many areas are also covered by thick deposits of windblown loess. The flatness of the landscape is due in large part to the thickness of these glacial deposits and the manner in which they were laid down (Fehrenbacher *et al.* 1968).

The deposits of a minimum of three major glaciations occur in Illinois. These include at least one pre-Illinoian glaciation (the Kansan), the Illinoian which was the most extensive glaciation in Illinois, and the Wisconsinan. Because the Wisconsinan was the most recent, its deposits overlie much of those representing the earlier glaciations. However, Illinoian-age deposits in areas outside the limits of Wisconsinan glaciation include some of the flattest topography in Illinois. At the same time, Illinoian deposits along drainages are heavily dissected; and on steep slopes where the loess cover has been eroded, well-developed paleosols are sometimes exposed (Fehrenbacher *et al.*1968).

Unglaciated portions of the state are characterized by rugged topography and bedrock outcrops of limestone, sandstone, and shale. Unglaciated areas occur only in extreme northwestern, western, and southern Illinois.

SOILS

Like topography, soils are of primary importance in the distribution of plant species. For example, soil particle size is reflected in drainage, aeration, water retention, and root penetration. Different types of parent materials produce soils with differing complements of various minerals and differing levels of natural fertility. Soil pH is largely dependent on parent material. Good soils combined with predominantly level topography and favorable climate make Illinois lands some of the most productive in the world.

Soil formation is dependent on the interrelationship between climate (particularly temperature and precipitation), soil parent material, soil organisms and plants, and time. Well-developed soils have readily recognizable zones produced by the translocation of minerals and soil particles. These factors have produced a multitude of soils; there are approximately 375 soil series in Illinois, grouped into 26 soil associations and five soil systems (mollisols, alfisols, inceptisols, entisols, and histosols) of which alfisols and mollisols are most important (Fehrenbacher *et al.* 1967).

The main soil parent materials in Illinois consist of loess, outwash, till, and alluvium (Fig. 4). The most extensively distributed is loess which covers approximately 64% of the state. Loess is silt that was blown off the floodplains of the major streams during the glacial periods and redeposited onto adjacent uplands. Because it was wind-transported, loess is thickest near the river valley source areas (Fig. 5). Loess-derived soils have relatively high natural fertility, friability, medium texture, and high water-retaining capacity (Fehrenbacher *et al.* 1968).

Soils derived from glacial till cover about 11% of the state and occur primarily in northeastern Illinois (Fig. 6). This area was the most recently glaciated of any region of Illinois and is far removed from the loess source areas along the Mississippi and Illinois rivers. Because the material being carried by the glaciers was not sorted prior to being deposited, the Wisconsinan tills are extremely variable in texture. Boulders are common in such deposits, as are sand, silt, and clay in various combinations. Medium-textured tills produce high-quality soils. The finer-textured tills rich in clay and the coarser-textured sands and gravels are lower in one or more factors such as natural fertility, water-retaining capacity or friability (Fehrenbacher *et al.* 1968).

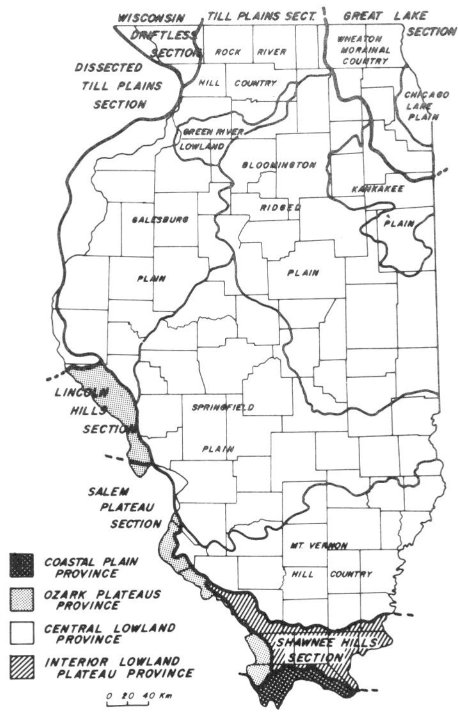

3. Physiographic divisions of Illinois (after Willman *et al.*, 1975).

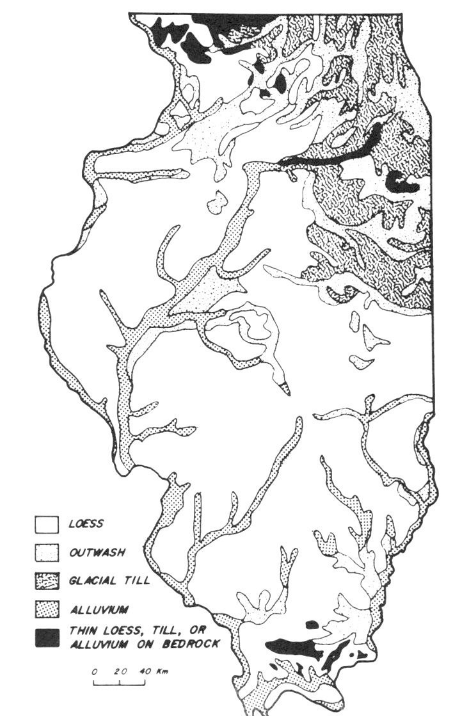

4. The main types of soil parent materials in Illinois (after Fehrenbacher *et al.*, 1967).

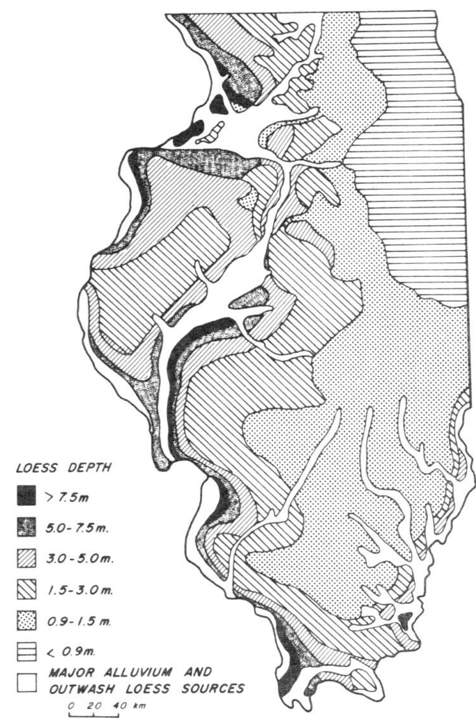

5. **Approximate loess depth (in inches) on uneroded, nearly level topography in Illinois (after Fehrenbacher *et al.*, 1967).**

Outwash materials form the basis for soils covering about 16% of the state. While materials deposited from glacial outwash vary in texture from gravel to clay, outwash sand deposits are by far the most extensive. They occur in northern Illinois and along the Mississippi, Illinois, Wabash, and Ohio river valleys. In the Mississippi and Illinois river valleys, large areas of outwash sand occur not only in the valleys but also on the adjacent uplands. The vegetation of the sand areas of Illinois is, in many ways, similar to that of forests and prairies in the more arid Great Plains to the west (Fehrenbacher *et al.* 1967).

Alluvium deposited by streams on their floodplains occurs throughout Illinois, occupying about 7% of the surface. Such deposits become extensive in the larger stream valleys. Alluvial sediments vary in texture from sands to clays depending on the source and the velocity of the water depositing them (Fehrenbacher *et al.* 1967).

Bedrock-derived soils are of minor occurrence (2% of the soils of the state). They are most important on steep slopes in the unglaciated sections of northwestern and southern Illinois. Organic soils formed from plant debris, such as peat, are also of minor importance, occurring primarily in old lake beds and marshes in northeastern Illinois.

In south-central and southern Illinois, the accumulation of clay in the subsoil has resulted in a relatively impermeable claypan limiting the downward penetration of both roots and moisture. As a result, the upper soil horizons may be saturated during one portion of the year and dry during another. Since roots are confined largely to the soil above the claypan, plants must be tolerant of both flooding and drought in order to exist in such an environment. This has resulted in a characteristic post oak or pin oak "flatwood" that occurs not only in south-central Illinois but wherever similar situations are found in the south-central and southern United states (Fehrenbacher *et al.* 1968, Schwegman *et al.* 1973).

Fragipan soils have accumulated silt horizons which, like the claypan, restrict air and water movement and downward root penetration. Although the available moisture of fragipan soils is high, plants are often unable to reach moisture in the lower horizons. Fragipan soils occur in thinner loess upland areas in south-central and southern Illinois (Fehrenbacher *et al.* 1968).

Soil Development

Although the parent materials of Illinois soils were primarily deposited during the Pleistocene, most of the soils themselves have developed during and since the Wisconsinan glaciation. Climate is an important factor in such soil development; it influences the composition of the vegetation and largely determines the type and rate of weathering occurring in the soil (Fehrenbacher *et al.* 1968). In general, both clay formation and degradation increase with increasing temperature and precipitation. The humid temperate climate that has characterized Illinois during the Holocene has, in part, hastened the breakdown of soil minerals as well as the formation and downward translocation of clay (Fehrenbacher *et al.* 1968).

Soils tend to be more heavily weathered and more highly developed in the thinner loess deposits farther from the source areas. Loess soils in Illinois are strongly developed only in south-central and southern Illinois where weathering occurred both during and after deposition of thin loess and where the underlying material may have accelerated soil formation (Fehrenbacher *et al*. 1968:168). Where the native vegetation was tall-grass prairie rather than forest, leaching of minerals and clay redepositon has not been serious enough to impair natural fertility or drainage.

At the present time, clay accumulation is occurring at a maximum rate in forested loess soils of west-central Illinois and at a minimum rate in northern Illinois because of lower temperature and lower precipitation. Clay degradation appears to be the most advanced in southern Illinois, especially in soils with fragipan and claypan horizons (Fehrenbacher *et al*. 1968:174-175).

PHYSIOGRAPHIC PROVINCES

The physiographic provinces defined by Leighton *et al*. (1948) are a series of regions with relatively constant geology, topography, and soils (Fig. 3). Approximately three-fourths of the state of Illinois lies in the Till Plains Section of the Central Lowland Province as defined by Leighton *et al*. (1948). The Till Plains Section is further divided into seven areas. Of these, four areas are covered primarily by Illinoian age glacial drift (Rock River Hill Country, Galesburg Plain, Springfield Plain, and Mt. Vernon Hill Country), two are lowlands with large areas of sand deposits (Green River Lowland, Kankakee Plain), and one is covered by Wisconsinan glacial deposits (Bloomington Ridged Plain). The Central Lowland Province also includes the Wisconsin Driftless Section, the Great Lakes Section (divided into the Wheaton Morainal Country and the Chicago Lake Plain), and a small area of the Dissected Till Plains Section extending across the Mississippi River from Missouri (Willman *et al*. 1975:7-19).

The Ozark Plateau Province includes the deeply dissected Lincoln Hills and Salem Plateau sections, portions of much more extensive landforms in Missouri. The Interior Low Plateau Province includes the Shawnee Hills Section, an unglaciated area of high topographic relief. The northern tip of the Coastal Plains Province extends into extreme southern Illinois. This province is characterized by rounded hills and extensive alluvial deposits (Willman *et al*. 1975:17-19). Physiographic provinces in large part coincide with

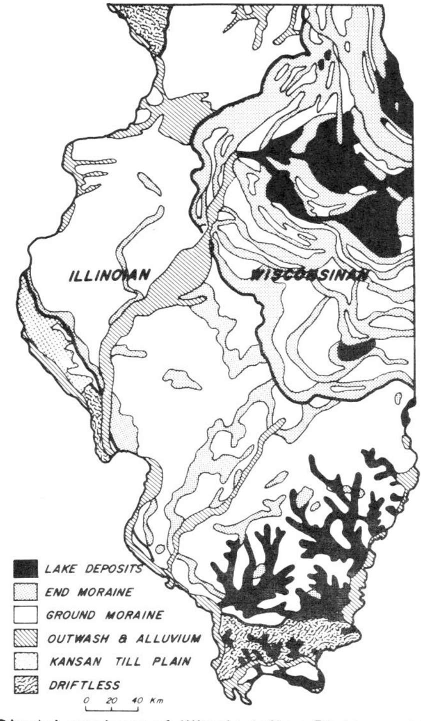

6. Glacial geology of Illinois (after Piskin and Bergstrom, 1967).

the Natural Divisions of Schwegman *et al*. (1973).

CLIMATE

The climate of Illinois is continental, characterized by hot summers and cold winters, although the cold of winter tends to be more extreme than the summer heat. Latitude is the principal control of both temperature and precipitation. Geographically, Illinois lies at the apex of a triangular semistationary mass of dry air extending eastward from the Rocky Mountains. The vegetation of this rather dry and cool region is dominated by grasses, thus earning the name "Prairie Peninsula." The northern border of the Prairie Peninsula, described by Transeau (1935), corresponds with the mean position of the January storm path and marks the southern edge of the zone of heavy snowfall. The southern border coincides with the mean winter position of the Gulf Coast Storm track and marks the average northern penetration of moist tropical air and heavy winter rainfall. Situated between these two boundaries, the central portion of Illinois is generally deficient in winter

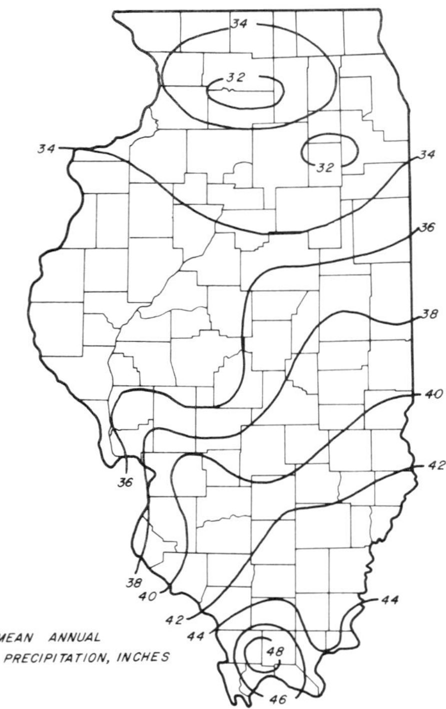

7. **Mean annual precipitation (in inches) for Illinois (after Denmark, 1974).**

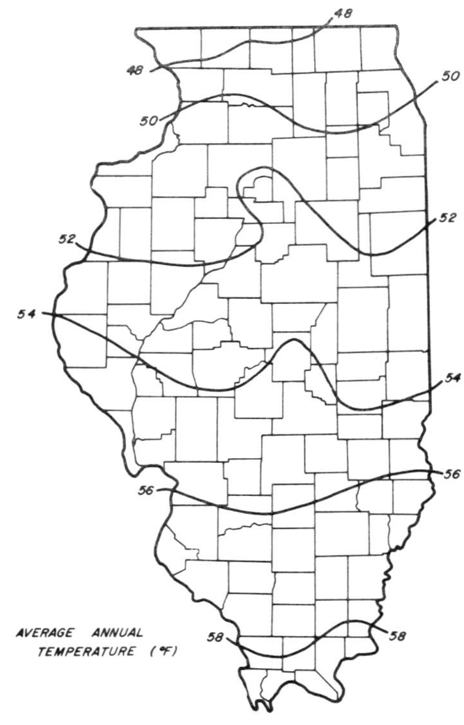

8. **Average annual temperature (degrees Fahrenheit) for the period 1931-1960 (after Fehrenbacher *et al.*, 1967).**

precipitation while the precipitation in successive months is relatively irregular and the annual and seasonal precipitation-evaporation ratios are lower than for areas outside the Prairie Peninsula (Transeau 1935). As a result, the area is markedly affected by the slight shortfalls in summer precipitation that often occur (Borchert 1950).

Mean annual precipitation (Fig. 7) varies from 48 inches in southern Illinois to less than 34 inches in the north. Average temperature (Fig. 8) ranges from 58 to 60° F in the south to 46 to 48° F in the north. The average length of the frost-free season varies from more than 200 days in extreme southern Illinois to less than 160 days for the north-central portion of the state (Fehrenbacher *et al*. 1967). The climate of Illinois is discussed in detail by Denmark (1974) and Page (1949).

Figure 9 is a climograph showing the average monthly precipitation and temperature for Rockford in northern Illinois, Springfield in central, and Cairo in extreme southern Illinois (based on data from Denmark 1974). A climograph is a way of graphically showing the relationship between temperature and precipitation. It is valuable to examine this relationship because vegetation is dependent on the interaction between the two factors, not just one or the other.

Not only is Cairo much warmer during the winter than the other stations but the winter precipitation is also approximately twice as high. Such winter precipitation is especially important for tree growth in the spring; it is one of the reasons that southern Illinois has more forest than the central portion of the state. Rockford, on the other hand, has relatively high precipitation during the late summer and early fall compared to Springfield or Cairo. Springfield, in the center of the Prairie Peninsula, receives little moisture in either the late summer or the winter. The bulk of the precipitation in central Illinois comes during the early summer. This early summer moisture is not critical for trees since they have already put on much of their annual growth using winter moisture retained in the soil. However, summer precipitation is critical for many warm-season grasses and agricultural crops, especially corn.

Figure 10 shows the average percentage of total precipitation occurring during each month for the three cities. The peak in central Illinois precipitation that occurs between April and June is readily apparent in

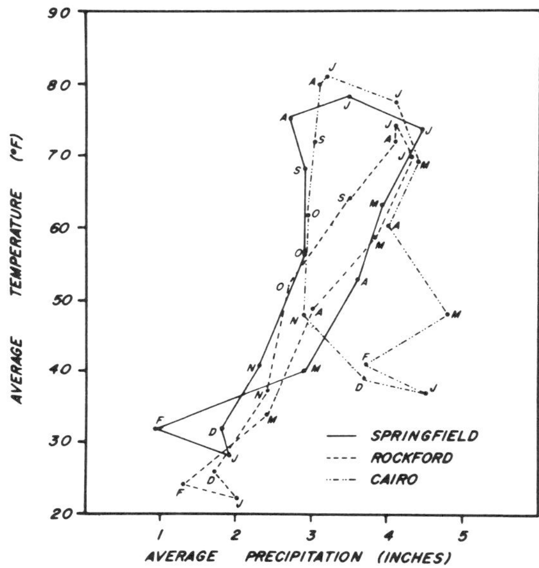

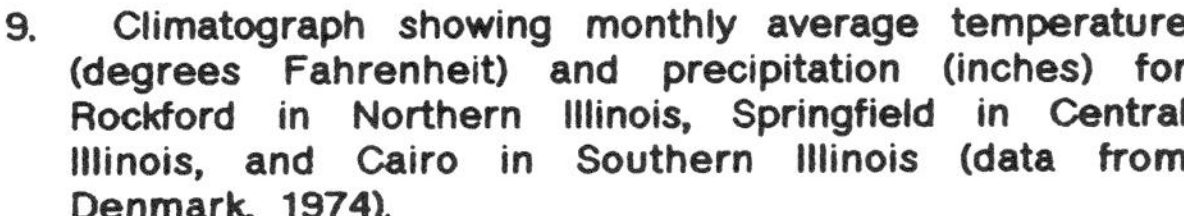
9. Climatograph showing monthly average temperature (degrees Fahrenheit) and precipitation (inches) for Rockford in Northern Illinois, Springfield in Central Illinois, and Cairo in Southern Illinois (data from Denmark, 1974).

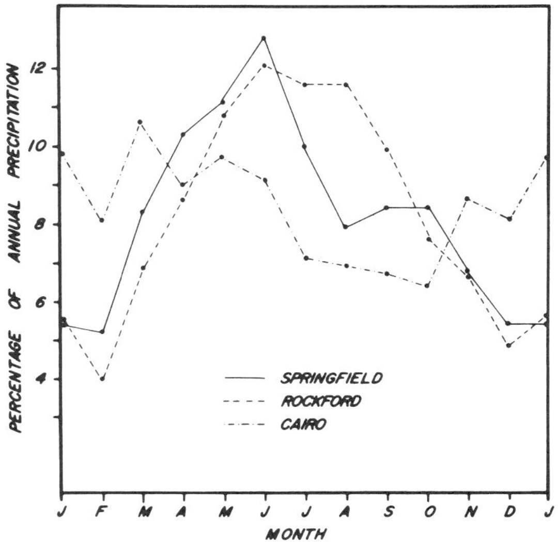

10. Graph showing the relative distribution of precipitation throughout the year for three Illinois cities (data from Denmark, 1974).

the data for Springfield. In northern Illinois, the precipitation is relatively high throughout the entire summer growing season. Although somewhat less precipitation occurs during the late summer than during other seasons at Cairo, the extremes of both wet and dry are much reduced.

VEGETATION AND THE NATURAL DIVISIONS

Because of the variation in climate, geology, and physiography across the state, a number of distinct biotic communities exist. Some of these communities are outliers of floristic or physiographic provinces that are more extensive in adjacent states. Such outliers include a small portion of the "Ozarks" in southwestern Illinois, the northern extension of the Coastal Plain forests into Illinois along the bottomlands of the Mississippi and Cache river systems, and the "Driftless Area" of northwestern Illinois. Other communities, most notably the Grand Prairie, reach their optimum development in Illinois.

The definition of natural divisions for Illinois (Schwegman *et al.* 1973) is an extremely valuable approach to natural environments and biotic communities and resources of the state. Schwegman *et al.* divided the state into 14 divisions containing 28 sections, on the basis of physiography, flora, and fauna (Fig. 11). The natural divisions are a synthesis of earlier work by ecologists such as Vestal (1931), Telford (1926), Braun (1950), Voigt and Mohlenbrock (1964) and P. Smith (1961). A series of "natural communities" has also been delineated for the state by White and Madany (1978) and is discussed in the Appendix.

The natural divisions of Illinois are summarized in Table 1 and discussed in greater detail by Schwegman *et al.* (1973) from whom the following brief descriptions are drawn.

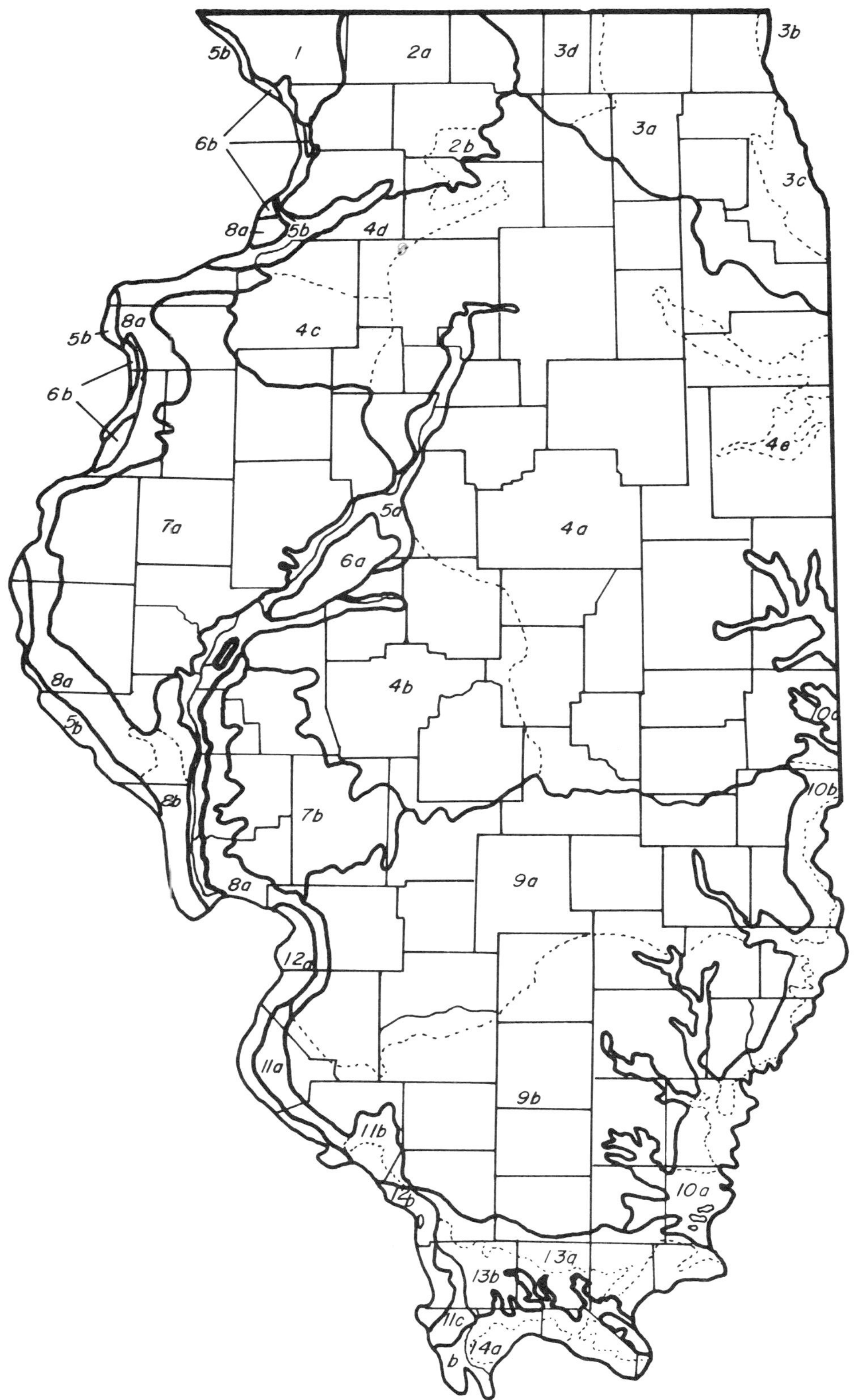

11. The Natural Divisions of Illinois (after Schwegman *et al.*, 1973).

1. The *Wisconsin Driftless Area* is part of an area extending into Wisconsin, Iowa, Minnesota, and Illinois that apparently escaped Pleistocene glaciation. The land surface is heavily dissected and bedrock outcrops and deep stream valleys are common. The rugged terrain was mostly forested at the time of Euro-American settlement.

2. The *Rock River Hill Country Division* is a region of rolling topography covered with a thin blanket of glacial till. At the time of Euro-American settlement, prairie covered most of the level uplands while forest occupied the stream valleys and the areas of broken topography.

3. The *Northeastern Morainal Division* was covered more recently by glaciers than any other part of Illinois. Glacial landforms are common and drainage is poorly developed. Forest and prairie both occurred in this division at the time of Euro-American settlement. There are also plant communities such as bogs which are characteristic of glacial landscapes.

4. The *Grand Prairie Division* is a flat, loess-covered plain that was originally characterized by tall-grass prairie with some forest on moraines and in stream valleys.

5. The *Upper Mississippi and Illinois River Bottomlands Division* is comprised of the rivers and floodplains of the Mississippi River above its confluence with the Missouri and the floodplains of the Illinois River and its major tributaries south of LaSalle. Much of the division was originally forested although prairie and marsh also occur.

6. The *Illinois River and Mississippi River Sand Areas Division* includes primarily the sand areas and dunes in the bottomland of the Illinois and Mississippi rivers. Scrub oak forest and dry sand prairie were the most common natural vegetation types in this division at the time of Euro-American settlement.

7. The *Western Forest-Prairie Division* is a strongly dissected glacial till plain. Prior to Euro-American settlement, the vegetation was chiefly forest although prairie frequently occurred on the level uplands.

8. The *Middle Mississippi Border Division* is a relatively narrow band of river bluffs and rugged terrain bordering the Mississippi River floodplain along much of Illinois. The original vegetation was predominantly forest with hill prairies occurring on west-facing bluffs.

9. The *Southern Till Plain Division* primarily includes the Illinoian till plain south of the limit of Wisconsinan glaciation. The topography is dissected, and the soils are generally poor because of the high clay content. Prior to Euro-American settlement, the vegetation consisted of both forest and prairie.

10. The *Wabash Border Division* includes the bottomlands and loess covered uplands of the Wabash, Vermilion, and Little Vermilion rivers and Crab Apple Creek. The area is characterized primarily by lowland oak forests containing many trees typical of forests to the east of Illinois.

11. The *Ozark Division* consists of the Illinois portion of the Salem Plateau which lies primarily in Missouri and northern Arkansas. The area is predominantly forested although hill prairies also occur.

12. The *Lower Mississippi River Bottomlands Division* includes the Mississippi floodplain from the confluence with the Missouri River south to Thebes Gorge. The northern section (the American Bottoms) originally contained prairies, marshes, and forest while the southern section was covered by a dense forest including some southern lowland species.

13. The *Shawnee Hills Division* is unglaciated hill country extending across southern Illinois and characterized by bedrock cliffs and hills. The division was originally forested.

14. The *Coastal Plain Division* is a region of swampy forested bottomlands and low hills. This region contains the northernmost extension of the Gulf Coastal Plain and is characterized by many southern plants and animals found nowhere else in Illinois.

TABLE 1. Summary of the Natural Divisions of Illinois, Based on Schwegman *et al*. (1973).

DIVISION	BEDROCK	SOILS	TOPOGRAPHY	PLANT COMMUNITIES
1. WISCONSIN DRIFTLESS AREA				
	Outcrops of limestone, dolomite, and shale along watercourses	loess or loess on bedrock	Mississippi River bluffs, interior stream canyons ravines, ridges, mounds, level to rolling upland	Forest: dry upland, mesic upland, floodplain Prairie: dry, mesic, loess hill Aquatic: rivers, creeks
2. ROCK RIVER HILL COUNTRY				
A. Freeport Section	Dolomite and limestone outcrops along streams	thick loess and thin loess on bedrock	level to rolling, river valleys, bluffs along streams, rolling uplands, meander scars	Forest: dry upland, mesic upland, floodplain Prairie: dry, mesic, wet Aquatic: marsh, rivers.
B. Oregon Section	sandstone outcrops common	thick loess and thin loess on bedrock	rugged topography with bluffs, ridges and ravines	Forest: dry upland, mesic upland, floodplain Prairie: dry, mesic, wet Aquatic: springs and seeps, rivers, creeks
3. NORTHEAST MORAINAL DIVISION				
A. Morainal Section	deeply buried by glacial drift, limestone outcrops along some streams	diverse soils, derived from glacial drift, lake bed sediments, beach deposits and peat	rolling upland, ravines, lake bluffs, glacial landforms (moraines, kames, eskers, drumlins, kettle-holes)	Forest: dry upland, mesic upland, floodplain tamarack swamp Prairie: dry, mesic, wet Aquatic: fen, marsh, sedge meadow, rivers, creeks, lakes, sloughs, bogs
B. Lake Michigan Sand Dunes Section	deeply buried by glacial drift	sandy with poorly developed soils	beaches, ridges, swales, and dunes	Forest: scrub-oak Prairie: dry sand, mesic sand, wet sand Aquatic: fen, marsh, creeks, Lake Michigan
C. Chicago Lake Plain Section	deeply buried by deposits of glacial Lake Chicago	lake bed sediments	flat, poorly drained	Forest: scrub-oak, mesic upland, floodplain Prairie: dry, mesic, wet Aquatic: fen, marsh, lakes, creeks, Lake Michigan
D. Winnebago Drift Section	deeply buried by glacial drift, outcrops	developed on early Wisconsinan glacial drift and loess	dunes, outwash plains, river terraces, meander scars	Forest: dry upland, mesic upland, floodplain Prairie: dry, mesic, wet, gravel hill, dry sand, mesic sand, wet sand, Aquatic: marsh, sedge meadow, rivers, creeks, sloughs
4. GRAND PRAIRIE DIVISION				
A. Grand Prairie Section	deeply buried outcrops along larger rivers	developed on thin to moderately thick loess, glacial drift or lakebed sediments, highly organic	glacial landforms common level to rolling upland, floodplains, ravines, river bluffs, and lake plain	Forest: floodplain, mesic upland dry upland, prairie groves Prairie: dry, mesic, wet Aquatic: marsh, prairie pot-holes, rivers, creeks
B. Springfield Section	deeply buried by Illinoian drift and loess, outcrops along Sangamon River	developed on loess and glacial drift, highly organic	level to rolling upland, floodplain, ravine, river bluffs	Forest: floodplain, mesic upland, dry upland Prairie: wet, mesic, dry, loess hill Aquatic: marsh, rivers, creeks
C. Western Section	deeply buried, occasional outcrops	developed on drift of Illinoian glaciation and loess	level to rolling upland, floodplain, ravines	Forest: floodplain, mesic upland, dry upland Prairie: wet, mesic, dry Aquatic: marsh, creeks
D. Green River Lowland Section	deeply buried	developed primarily on sand deposited as glacial outwash	outwash plain, dunes	Forest: scrub-oak, floodplain Prairie: wet, mesic, wet sand, mesic sand, dry sand Aquatic: marsh, rivers

TABLE 1. (Continued).

DIVISION	BEDROCK	SOILS	TOPOGRAPHY	PLANT COMMUNITIES
E. Kankakee Sand Area Section	deeply buried	developed primarily from sand, deposited by Kankakee flood during Late Wisconsinan glaciation	outwash plain, dunes	Forest: scrub oak Prairie: wet sand, mesic sand, dry sand Aquatic: marsh, creeks
5. UPPER MISSISSIPPI RIVER AND ILLINOIS RIVER BOTTOMLANDS DIVISION				
A. Illinois River Section	deeply buried by alluvium	soils formed on recent alluvium and glacial outwash	broad floodplains ad terraces formed by glacial meltwater	Forest: bottomland Prairie: wet, mesic Aquatic: marsh, spring bogs, backwater and oxbow lakes, rivers
B. Mississippi River Section	deeply buried by alluvium	soils formed on recent alluvium and glacial ouwash	broad floodplains terraces formed by glacial meltwater	Forest: bottomland Prairie: wet, mesic Aquatic: marsh, oxbow lakes, rivers
6. ILLINIOIS RIVER AND MISSISSIPPI RIVER SAND AREAS				
A. Illinois River Section	deeply buried by water or wind laid sand deposits	developed from sandy materials in bottomland of Illinois River	dunes, blowouts, level to rolling plain	Forest: scrub-oak Prairie: dry sand, mesic sand, wet sand Aquatic: marsh
B. Mississippi River Section	deeply buried by water or wind-laid sand deposits	developed from sand and sandy materials in bottomland of Mississippi River including "perched dunes" atop bluffs in Jo Daviess Co.	dunes, blowouts, level rolling plain	Forest: scrub-oak Prairie: dry sand, mesic sand, wet sand Aquatic: marsh
7. WESTERN FOREST-PRAIRIE DIVISION				
A. Galesburg Section	buried by Illinoian and pre-Illinoian age glacial drift, outcrops along major streams	fairly young soils developed on 4-5 feet of loess, highly organic	strongly dissected till plain, ravines, level to rolling uplands, floodplains	Forest: dry upland, mesic upland, floodplain Prairie: dry, mesic, wet Aquatic: marsh, rivers, creeks
B. Carlinville Section	buried by Illinoian and pre-Illinoian age glacial drift, outcrops along major streams	fairly young, developed on 4-5 feet of loess, highly organic	strongly dissected till plain, ravines, level to rolling	Forest: dry upland, mesic upland, floodplain Prairie: dry, mesic, wet Aquatic: marsh, creeks
8. MIDDLE MISSISSIPPI BORDER DIVISION				
A. Glaciated Section	limestone, sandstone, and dolomite outcrops common along bluffs	soils on uplands developed from deep, well drained loess	heavily dissected, river bluffs and rugged terrain bordering the Mississippi River	Forest: dry upland, mesic upland, floodplain Prairie: loess hill Aquatic: creeks
B. Driftless Area	limestone and sandstone outcrops	soils on uplands developed from deep, well-drained loess	heavily dissected, river bluffs, ravines, floodplain, sinkholes, sinkhole plain	Forest: dry upland, mesic upland, floodplain Prairie: loess hill Aquatic: creeks, sinkhole ponds
9. SOUTHERN TILL PLAIN DIVISION				
A. Effingham Plain Section	sandstone, limestone, coal, and shale outcrops	developed on thin loess and Illinoian glacial till plain, high clay content, fragipan and claypan layers characteristic of upland soils	nearly level to dissected till plain, broad floodplains along major streams, ravines	Forest: upland flatwoods, dry upland, mesic upland, floodplain Prairie: wet, mesic, dry Aquatic: marsh, creeks, rivers, oxbow lakes

TABLE 1. (Continued).

DIVISION	BEDROCK	SOILS	TOPOGRAPHY	PLANT COMMUNITIES
B. Mt. Vernon Hill Country Section	sandstone, limestone, coal, and shale near surface, outcrops	developed on thin loess and Illinoian glacial till plain, high clay content, fragipan and claypan layers characteristic of upland soils	hilly and rolling, broad floodplains along major streams, ravines	Forest: upland flatwoods, dry upland, mesic upland, floodplain Prairie: wet, mesic, dry Aquatic: marsh, creeks rivers, oxbow lakes
10. WABASH BORDER DIVISION				
A. Bottomlands Section	covered by Illinoian till, small outcrops of sandstone, limestone, coal, and shale along some streams	floodplain and terrace soils	river floodplain, terrace, meander scars	Forest: floodplain, terrace, swamp Prairie: mesic, wet Aquatic: marsh, oxbow lakes, sloughs, rivers
B. Southern Uplands Section	covered by Illinoian till, sandstone outcrops	developed from moderately deep loess on uplands and bluffs	dissected till plain, river bluffs, ravines	Forest: dry upland, mesic upland, floodplain Prairie: mesic, dry Aquatic: creeks
C. Vermilion River Section	buried by Wisconsinan loess	developed from thin loes over loamy till	rugged topography, till plain, ravines, floodplain	Forest: dry upland, mesic upland, floodplain Prairie: dry, mesic, wet Aquatic: rivers, creeks
11. OZARK DIVISION				
A. Northern Section	Illinois portion of Salem plateau, limestone bedrock and outcrops	derived from deep loess, with thin soil on bedrock outcrops along bluffs and in ravines	well developed sinkhole plain, ravines, river bluffs, floodplain, maturely dissected plateau	Forest: dry upland, mesic upland, floodplain glade Prairie: loess hill Aquatic: sinkhole ponds, creeks, springs
B. Central Section	Illinois portion of Salem plateau, sandstone outcrops and bedrock	derived from deep loess, thin soils on bedrock outcrops along bluffs and in ravines	hills, ravines, stream canyons, floodplain, maturely dissected plateau	Forest: dry upland, mesic upland, floodplain Aquatic: creeks
C. Southern Section	Illinois portion of Salem plateau, cherty limestone bedrock and outcrops	derived from deep loess, with thin soils on bedrock outcrops along bluffs	steep ravines, river bluffs, floodplain maturely dissected	Forest: dry upland, mesic upland, floodplain Prairie: loess hill Aquatic: creeks, springs, sinkhole ponds
12. LOWER MISSISSIPPI RIVER BOTTOMLANDS DIVISION				
A. Northern Section (American Bottoms)	deeply buried by alluvium	developed from alluvium, generally fine-textured, with areas of both sandy, well-drained soil and clay soils with poor internal drainage	broad bottomlands formed by glacial waters	Forest: bottomland Prairie: wet, mesic Aquatic: marsh, oxbow lakes, Mississippi River
B. Southern Section	deeply buried by alluvium	developed from alluvium, generally fine-textured, areas of both sand and clay	broad bottomlands formed by glacial waters	Forest: bottomland on heavy soil, bottomland on light soil, bottomland swamp Aquatic: oxbow lakes, Mississippi River, springfed swamp
13. SHAWNEE HILLS DIVISION				
A. Greater Shawnee Hills Section	sandstone bedrock and cliffs	derived mainly from loess, most strongly developed on thin loess,	very rugged, many bluffs and ravines; north slopes relatively gentle, south slopes with escarpments, cliffs, and overhanging bluffs,	Forest: dry upland, mesic ravine, floodplain Aquatic: creeks

Table 1. Concluded.

DIVISION	BEDROCK	SOILS	TOPOGRAPHY	PLANT COMMUNITIES
B. Lesser Shawnee Hills Section	limestone and sandstone bedrock and outcrops	derived mainly from loess, most strongly developed on thin loess, claypan and fragipan soils common	rugged with many bluffs and ravines, gentler than Greater Shawnee Hills, sinkhole topography	Forest: dry upland, mesic ravine, floodplain Prairie: limestone glade Aquatic: sinkhole ponds, creeks
14. COASTAL PLAIN DIVISION				
A. Cretaceous Hills Section	unconsolidated Cretaceous and Tertiary sands gravels, and clays, gravel exposures	upland soils derived from relatively thin loess or gravel	steep to rolling hills capped by Cretaceous and Tertiary sand, gravel, and clay	Forest: dry upland, mesic upland Prairie: dry, mesic Aquatic: southern seep spring, creeks
B. Bottomlands Section	deeply buried by alluvium	soils developed on recent alluvium and older terrace deposits which are generally clays	broad floodplain at confluence of the Ohio and Mississippi rivers and the broad floodplain of the Cache; terraces and meander scars	Forest: floodplain, terrace, swamp Aquatic: backwater swamps, sloughs, oxbow lakes, creeks, rivers

FLORISTIC PROVINCES

Anderson and Ugent (1980) defined a series of floristic provinces for the state based entirely on plant distribution rather than on the combination of flora, fauna, and geology used to define the natural divisions. The recognition of such floristic provinces is based on the fact that regions with pronounced differences in climate or soils also frequently show relatively abrupt spatial changes in plant species composition that are in turn reflected in an increase in the number of range terminations. Plant communities in areas of high species turnover are usually transitional between those occurring in the areas on either side where environmental variability is not as great. Such transitional areas were termed "tension zones" by Curtis (1959) since rather small environmental changes might shift the vegetation toward one vegetation type or the other. Anderson and Ugent (1980) identified four floristic provinces for Illinois (Fig. 12). These include the Northeast Forest, Grand Prairie, Prairie-Forest, and the Southern-Forest. Three transition or "tension zones" separate the floristic provinces.

The Northeast Forest Province contains many northern forest species that become restricted in distribution farther south. Many of these plants are common boreal or northern forest taxa.

The Grand Prairie and Forest-Prairie Provinces are the least distinct, with no sharp boundary separating the two. Prior to Euro-American settlement, the Grand Prairie Province was a region with broad, open, prairies; in the Forest-Prairie Province, the prairies were extensively dissected by forests (Anderson and Ugent 1980, Anderson 1970).

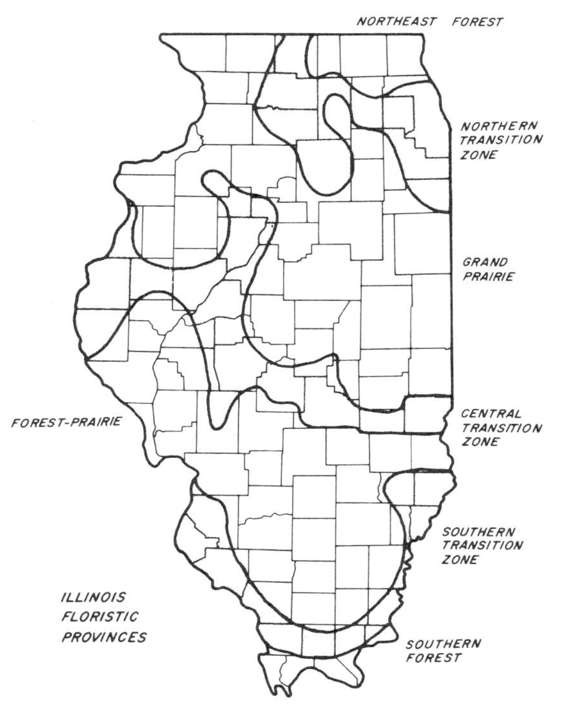

12. The distribution of floristic provinces in Illinois (based on Anderson and Ugent 1980).

The Central Floristic Transitional Zone reflects the intersection of the southern Ozarkian forest elements with the Grand Prairie. The northward extension of this transitional zone along the Illinois River includes sand deposits and their characteristic flora with southern Illinois species extending up the Illinois River and northern species extending southward down the river. Not only does the Illinois River Valley provide protection from sun and wind desiccation for mesic species but the steep bluffs and slopes offer bedrock outcrops and dry, west-facing exposures for xeric plants as well. Thus, in the Illinois River Valley, the species characteristic of cooler, more mesic forests to the north and east meet with those characteristic of the drier forests to the south and west.

EFFECTS OF CLIMATE ON VEGETATION

It is the climatic extremes rather than the means that control the survival and reproduction of plants. Temperature extremes occur as either severe winter cold or excessive summer heat, each generally affecting different plants. In addition, the length of the growing season is also a factor, determined by the last killing frost in the spring and the first in the fall.

Extremes in precipitation are reflected not only in the total amount but also in seasonal distribution and the frequency and intensity of storms. Herbaceous plants rely on precipitation that occurs during the growing season while tree growth is based primarily on precipitation which fell during the previous winter.

Although temperature and precipitation are often reported as average annual figures, it is really the monthly or short-term values that are the most significant for plant growth. The result of interaction between temperature and precipitation, known as effective precipitation, is critical to plants. For example, when it is cool, less evaporation takes place, a plant's water needs are reduced, and a larger amount of moisture remains available.

An illustration of how meaningless average temperature and precipitation figures can be is shown by the comparison of climatic data from Springfield, Illinois (Denmark 1974), and Seattle, Washington (Blair and Fite 1965). Springfield, lying in the center of the Prairie Peninsula, has an average annual temperature of 53.6° F and an average annual rainfall of 34.8 inches. Seattle, which lies in the Cedar-Hemlock-Douglas Fir forest of the Pacific Northwest, is slightly cooler and actually somewhat drier with an average annual temperature of 51.0° F and rainfall of 34.03 inches. However, in the Pacific Northwest, 80% of the rainfall occurs (often as slow drizzle) between October and March while in central Illinois only about 40% of the precipitation occurs during this period. In addition, the average monthly temperatures for Seattle range from 39.5° F in January to 63.1° F in July and August (a range of 23.6° F) while the temperatures for Springfield range from 28.4° F in January to 78.0° F in July (a range of 49.6° F). The range of average monthly temperatures for Springfield is more than twice that of Seattle where the winters are comparatively warm and rainy and the summers cool and dry.

Effects of Temperature

Some plants, generally those of southern affinities, can withstand heat well but are susceptible to frost. Other species are tolerant of unusually cold conditions while wilting rapidly when too hot. Some plants can withstand heat provided adequate moisture is available while others actually require a hot, dry environment.

As far as potential food production is concerned, cold temperature is a problem when it occurs near the time of flowering or before fruit maturation. Once destroyed, flower buds or young fruit will not be reformed that year. Sometimes temperatures drop well below freezing and no plants escape injury; at other times, frost is slight and plants in sheltered locations are unaffected. Cold air, being heavier than warm air, sinks, flowing down valleys and accumulating in low places. Therefore, the most susceptible trees are generally those that bloom early when the likelihood of freezing is greatest or those that grow in the moist soils of valley bottomlands where cold air is likely to settle. Several bottomland oaks as well as pawpaw and persimmon are likely to be damaged by freezing as are modern fruit trees such as apples and peaches and many other species planted north of their normal range. On the other hand, the habitats most prone to freezing often are least affected by extreme heat during a summer drought.

Effects of Precipitation

Depending on the severity of drought, plants are affected in several ways including reductions in growth, vigor, and yield and even ultimately death. Fruits of many trees fail to develop during periods of too little moisture. The first trees hit by drought are those growing on ridges and exposed slopes and those on rocky or sandy soils or soils with a fragipan or claypan layer limiting root penetration. Trees on deeper soils or on protected slopes and those in river valleys where there is deep soil moisture are the least likely to be affected by short-term drought. However, during droughts of long duration, such as occurred in the Midwest during the 1930's, trees in moist ravines where there was no permanent water died much sooner than those more deeply rooted in drier upland soils or where permanent water was available (Albertson and Weaver 1945: 414,431).

The chief causes of tree mortality during severe droughts are lack of water in combination with competition from grasses, decreased water infiltration and rapid runoff, drying up of springs and streams, a fall in the water table in ravines and lowland terraces, low humidity and high evaporation, desiccating winds, and the inability of trees to accommodate their root systems to the rapidly changing environment. However, where a dry season is a regular and characteristic feature of the climate, no matter how severe, the native plants are so well adapted that they generally show no ill effects. Drought is common in all climates, but in moist climates a relatively short rainless period will be as damaging as a much longer period in a semiarid environment (Daubenmire 1959: 103).

Alternatively, too much moisture can also be harmful. Only a few floodplain species can tolerate their roots being submerged for more than a brief period. The pollination of some species, such as pecan, is adversely affected by rainfall during the flowering period.

Relationship Between Soils And Drought Susceptibility

The productivities of soils in different portions of Illinois are affected by various limiting factors (Fig. 13, Fehrenbacher *et al.* 1968). Much of southern Illinois, for example, has fragipan or claypan soils or shallow soils underlain by bedrock. This is one of the primary reasons that southern Illinois has an oak-hickory dominated forest rather than a forest with a more mesic composition.

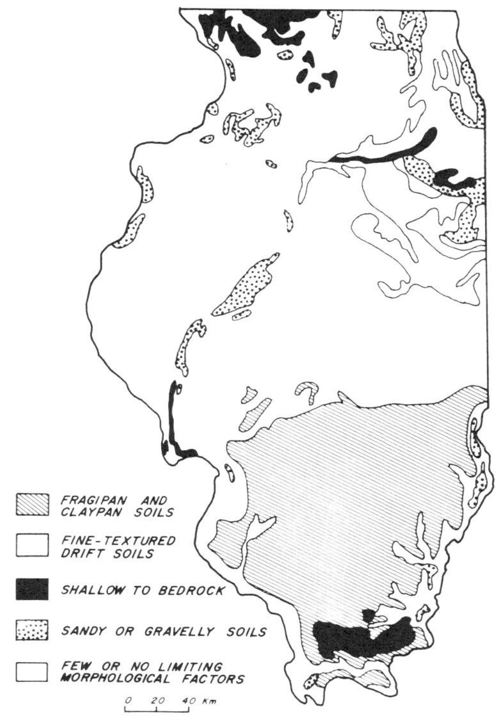

13. Soil morphological features limiting crop production in Illinois (after Fehrenbacher *et al*, 1968).

There is strong evidence that the soil-forming factors may change over a relatively brief period. Recent studies indicate that climate, vegetation, erosion cycles, geomorphic surfaces, and associated changes in parent materials, slope, and soil moisture have changed considerably since the deposition of the parent materials (Fehrenbacher *et al*. 1968: 170).

Two characteristics of young versus more mature soils are important to the understanding of vegetational patterns. First, younger soils generally are higher in available nutrients and, under comparable conditions, are superior in crop-producing potential (Buckman and Brady 1969: 292). Second, effects of weathering are cumulative and therefore less well expressed in young soils.

Parts of northern Illinois have shallow bedrock and soils derived from sand, gravel, and fine-textured glacial drift. These soils have relatively low water-storage capacities; and unless precipitation is abundant, productivity is low. Soils of most of the remainder of the state have high water-storage capacities and natural fertility. However, much of central Illinois coincides with the apex of the Prairie Peninsula; consequently, climatic restraints rather than edaphic ones become the factors limiting plant growth. The climate of this region is not conducive to tree growth because precipitation is sporadic and occurs primarily during the summer while most tree growth is on the basis of winter precipitation. However, the precipitation regime is ideal for the growth of grasses and herbaceous plants and, combined with the high quality soils and level topography, was responsible for the development of the tall-grass prairie and its persistence for the last 8000 years.

The early Holocene landscape was undoubtedly undergoing climatically induced modifications, such as erosion on slopes and uplands and redeposition in floodplains. Despite such changes, the relatively uniform mantle of unweathered loess probably made the vegetation of Illinois considerably less variable than it is today. Relatively little change has occurred during most of the Holocene in the natural fertility of those loess soils on level uplands that have been covered by prairie. This is primarily the result of increasing organic matter, decreased leaching, and very low rates of erosion. On the other hand, the thin loess soils of southern Illinois that were primarily forested are strongly developed and frequently characterized by claypan and fragipan accumulations. Thick loess soils in western Illinois that support forests are currently undergoing clay formation and translocation of minerals at a relatively high rate compared to northern or eastern Illinois (Fehrenbacher *et al.* 1967). Thus, soil development along with decreasing natural fertility and internal drainage have not occurred at the same rate throughout Illinois.

At some time in the past, the difference in soil development in various regions of Illinois became significant in plant growth. Ultimately, the vegetational differences resulting from those soil changes became great enough to affect human subsistence patterns. Unfortunately, soil development was both influenced and masked by the effects of climatic change during the Holocene. An oak-hickory forest existing because soil conditions limit water availability may be difficult to distinguish from one resulting from reduced precipitation. The reason a forest exists may be unimportant in terms of human food-resource availability. However, many other factors such as the availability of water in streams and springs are relatively unaffected by soil formation but heavily affected by the precipitation regime.

PRESETTLEMENT VEGETATION

Although the aboriginal inhabitants of Illinois did influence the environment to a limited extent, the term "presettlement" is used here to refer to the period preceding Euro-American settlement and its subsequent radical alteration of the landscape. Once it was realized that the prairie was not a kind of desert but rather a very rich and valuable type of cropland, the settlement of Illinois by Euro-Americans proceeded swiftly. Prior to being sold to settlers, land was surveyed by the General Land Office; in Illinois, these surveys took place between 1813 and 1849. The GLO surveyors established township, section, and quarter-section points. Surveyors were instructed to blaze two trees (or four in later surveys) in opposite directions at each point with survey data. The distance, bearing, species, and diameter of these trees were recorded as an aid in their relocation. These historic data form essentially a random sampling of the presettlement forests. In addition, the surveyors recorded trees that occurred on the survey lines, homesteads and other improvements, bodies of water, evidence of fires and natural catastrophes, and any notable features. At the end of each mile, the surveyors evaluated the quality of the forest and soils and rated the suitability of the land for settlement and cultivation (Stewart 1935). Copies of the survey notes and plat maps are available from the State of Illinois, Division of Archives and Records, Springfield.

Because Euro-American settlement was so rapid, occurring in only one or two decades, the General Land Office Survey records are the only data on the original early historic vegetation available in many areas. Numerous techniques have been developed for using these data to reconstruct past forest structure and composition (Cottam and Curtis 1949, 1955, 1956; Cottam *et al.* 1953) and to minimize possible bias and error (Bourdo 1956, F. King 1978, Wood 1976).

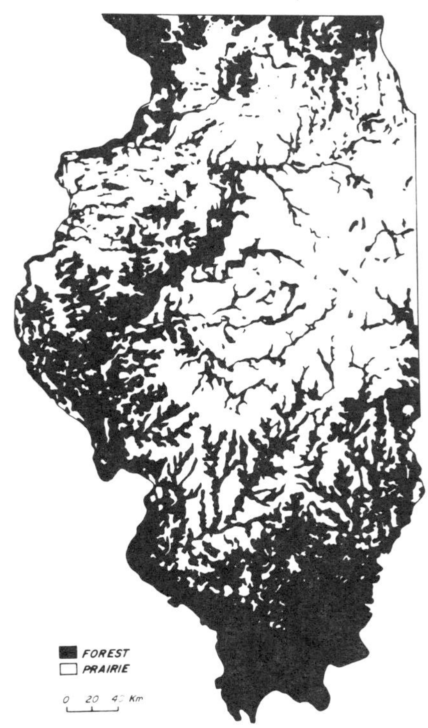

14. Distribution of forest and prairie in Illinois about 1820, based on General Land Office survey data (after Anderson 1970).

The distribution of prairie and forest (Fig. 14) based on the General Land Office data (Anderson 1970) indicates that, outside of the Grand Prairie Division, much of the state was covered by a patchy mosaic of forest and prairie. Prairie often covered the uplands while forest generally followed the stream valleys. Because summer winds most frequently come out of the southwest, the eastern and northern sides of stream valleys were often more heavily forested than the western sides (Anderson 1970, Gleason 1913, F. King and Johnson 1977).

A number of authors have discussed the presettlement vegetation of Illinois on the basis of the General Land Office survey data. Anderson (1970) discussed the distribution of forest and prairie. Studies of the presettlement vegetation have been published for Williamson County (Anderson and Anderson 1975), unglaciated Southern Illinois (Leitner and Jackson 1981), the Sangamon River Basin (F. King and Johnson 1977), McLean and Mason counties (Rodgers and Anderson 1979), Jo Daviess County (M. Jackson 1977), and Lake County (Moran, 1976).

PAST CLIMATES AND VEGETATION

Because of the geographical location of the state and its unique combination of glacial history, topography, and climate, prairie has dominated the landscape of Illinois for thousands of years. Illinois was glaciated at least four times during the Pleistocene. The ice of the third, or Illinoian, glaciation was the most extensive, covering approximately 90% of the state. As recently as 18,000 years ago, one-third of Illinois was buried under ice of the most recent (Wisconsinan) glaciation. The remainder of the state was covered by barren ground and boreal forest. After 18,000 YBP (years before present) the climate began to warm, the ice started to melt, and the margins of the ice-sheet retreated northward. Because of the long north-south extent of the state, climatic changes associated with the retreat of the Wisconsinan glaciers were time-transgressive, proceeding from south to north over a period of many thousand years. The last glacial ice retreated from northeastern Illinois sometime around 14,000 years ago.

There have been several reconstructions of the postglacial environment of Illinois based on fossil pollen evidence (Gruger 1972, J. King *et al.* 1976, C. Griffin 1951, Peterson 1976, Voss 1937, Wright 1968). Most recently J. King (1981) presents a transect across the length of the state providing a vegetation record for most of the last 14,000 years.

As the glaciers withdrew, the late-glacial vegetation of Illinois was dominated by spruce forest in the north and open spruce woodland in central Illinois. These communities disappeared in a northward direction between 14,000 and 10,900 YBP. In central Illinois, these forests were followed by ash in the lowlands while the uplands apparently remained treeless. In northern Illinois spruce was replaced by a pine-ash forest. Both the ash of central Illinois and the pine-ash of northern Illinois were subsequently replaced by a mixture of cool-temperate and then warm-temperate tree species. By 9000 years ago, all of Illinois was dominated by deciduous forest (Fig. 15) (J. King 1981).

Prairie development started between 8500 and 7900 YBP, occurring first along the southern margin of the Prairie Peninsula. This marked the start of the "Hypsithermal" climatic interval and is also correlated with

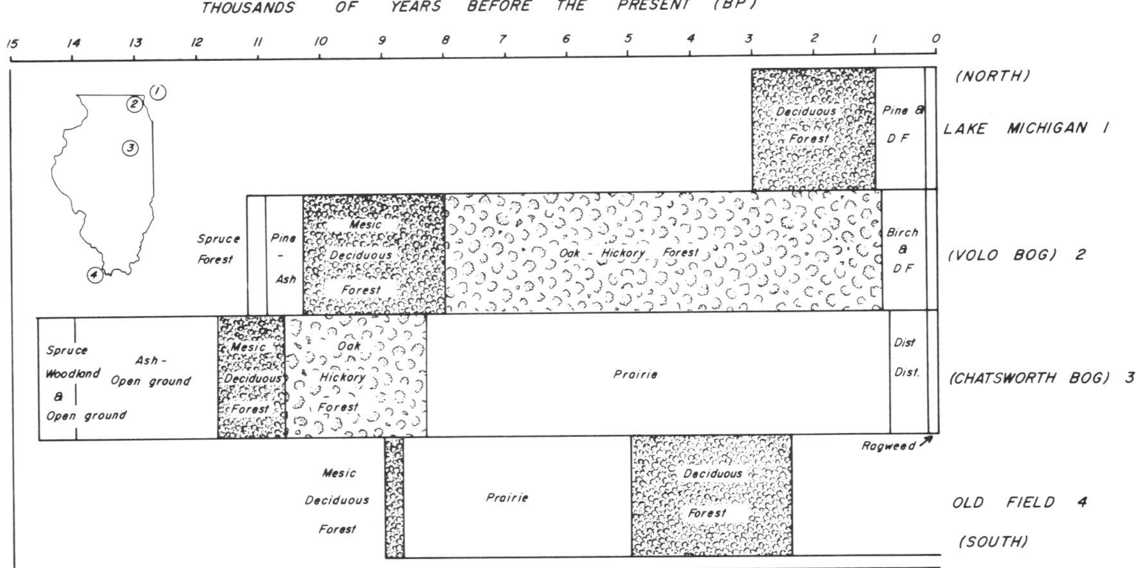

15. **Distribution of vegetation types in Illinois through time, based on pollen evidence (after J. King, 1981).**

the final decay of the Laurentide ice sheet and the establishment of the modern system of atmospheric circulation over North America (J. King 1981). In central Illinois, prairie vegetation began encroaching on forest about 8300 YBP in response to these warmer and/or drier conditions of the Hypsithermal. By 7900 YBP, dry oak-hickory forest had replaced the more mesic forests of northern Illinois as well.

About 5500 YBP, the southern border of the Prairie Peninsula began to experience somewhat increased moisture with resultant forest expansion. However, lying in the middle of the Prairie Peninsula, central Illinois has remained grassland to the present. While not as drastically affected by the Hypsithermal, the forests of northern Illinois never returned to the more mesic deciduous forests of the early Holocene (J. King 1981).

Between 900 and about 200 years ago, Midwestern climate apparently became cooler once again. This change is correlated with late stages of the "Neoglacial" or "Little Ice Age," a period of increasing cold and advancing mountain glaciers. During this period, cool temperate species such as pine and larch returned to northern Illinois. This period may also coincide with forest expansion in central Illinois and the ultimate formation of modern prairie groves (Geis and Boggess 1970, J. King 1981).

As stated by Anderson (1970:216), "Illinois has been the battleground between forest and prairie for thousands of years." The boundary between the two has shifted continually in response to climatic shifts which favored one or the other. For example, vegetation maps based on soils (Vestal 1931) show less prairie than maps based on data from General Land Office surveys conducted in the early 1800's (Anderson 1970).

Temperature and precipitation have been particularly important to the development of Illinois vegetation and therefore to human subsistence. During the past, not only have the average temperature and precipitation varied but also the extremes of temperature and precipitation and the seasonal distribution and intensity of rainfall. Wendland and Bryson (1974) suggest that precipitation during the Holocene in Illinois has varied by as much as 20% from the present and temperature by 5 to 6° F for long periods of time.

Climatic oscillations since the Pleistocene have followed a trend toward increasing temperatures culminating in the Hypsithermal (approximately 8500-5500 years ago) followed by cooling. Prior to 10,000 years ago, the growing season was probably 1 to 2 months shorter than today. During the Hypsithermal the growing season may have lengthened to about a month longer than it is currently (Wendland and Bryson 1974). Although there have been a number of fluctuations, the climate gradually started cooling about 5000 years ago (J. King 1981).

WILD PLANT FOODS: DISTRIBUTION AND RELIABILITY

The use of wild plants in present-day America is minimal compared to that of aboriginal societies. While most people probably would not hesitate to collect fruit or nuts belonging to familiar genera, such as strawberries or black walnuts, there is enough suspicion of many other less familiar but supposedly edible plants that few people except experts or back-to-nature enthusiasts collect them for food. This is especially true for those plants that also have poisonous parts, require special preparation techniques, or are difficult to distinguish from closely related poisonous species.

Many outstanding wild plant foods have been domesticated and can be grown in the home garden. In many cases, wild plant foods that are potentially economically profitable are eventually cultivated. For example, wild rice has brought a high price for many years because it did not adapt well to commercial pond cultivation and only a limited amount of the wild grain was available. As a result of the high price, however, great effort has been put into attempts at cultivation, and the techniques to pond-raise wild rice have been perfected.

Most current experimentation with wild plant foods is casual. Aside from wild fruits and nuts, the greatest use of wild plants is probably as flavorings (mints, wintergreen, wild onions), in salads or as cooked greens, or for use as herbal teas. The plants that have remained entirely "wild" generally have done so because they are in some way unappealing and/or unable to compete with other foods, usually of Old World origin, for a place in our modern diet. Acorns, for example, in addition to frequently containing insect larvae also contain bitter tannins which must be laboriously leached out prior to using the nutmeats. Groundnut (*Apios americana*), known to early settlers as pomme-de-terre, was an important food to Indians and early settlers alike; today it is completely ignored. Other edible native tubers which are largely ignored today include *Sagittaria latifolia* (duck potato or arrowhead), *Cyperus esculentus* (nut-sedge), *Nelumbo lutea* (American lotus), *Nuphar luteum* ssp. *macrophyllum* (yellow pond-lily), *Heracleum maximum* (cow parsnip), and several others. Although a related species of lotus is used in the Orient for food, none of these or any other of the wild tubers used by the North American Indians are either cultivated or commonly collected today. The obvious reason is that the introduced *Solanum tuberosum* (white potato) and *Ipomoea batatas* (sweet potato), both of which produce larger, more easily harvested tubers, have essentially replaced those of wild plants. A possible exception is *Helianthus tuberosus* (Jerusalem artichoke or sunchoke), but even that is sold more as an exotic than as a staple. (A popular modern cookbook lists one recipe for Jerusalem artichokes and 58 for potatoes.) Although many rural folk continue to collect "greens" for potherbs in the spring, most are probably considered less desirable than cultivated types such as lettuce or spinach.

The number of cultivated plant foods available today is great. The average grocery store stocks hundreds of plant foods in one form or another, fresh, frozen, canned or dried. Thus, there is little motivation to collect wild foods, and many people lack the time or the opportunity in any case. Prehistoric peoples, on the other hand, were more dependent on the natural environment. Prior to the development of horticulture, Indians of eastern North America had to rely on what native foods they could catch or collect.

TYPES OF PLANT FOODS

Those portions of plants used by people for food generally are produced by the plant for another purpose: either reproduction (flowers, seeds, acorns, nuts), nutrient movement (sap), growth (cambium), food storage (roots, bulbs, tubers), or food production (leaves). In most cases, the only plant parts which have evolved to be eaten are fleshy fruits and berries as an aid in the distribution of seeds by birds and animals.

It is to the advantage of a plant species to disseminate its propagules as widely as possible. Some plants do this by producing light, often winged, seeds which can be carried great distances by the wind. Other plants have seeds enclosed in fleshy edible fruit which are eaten by animals and carried miles before being eliminated. Frequently, plant species with edible fruit have evolved hard, thick seed coats which must pass through the digestive tract of some animal and be etched by digestive acids before germination can occur. This mechanism guarantees that some of the seeds will be deposited far from where they were eaten. Of these, some will be dropped in sites favorable for germination and growth even though the majority may perish.

TABLE 2. Comparison of the Composition of 100 grams of Various Plant Foods

FOOD	FOOD cal.	VALUE cal.*	WATER %	PROTEIN	FAT	CARBO-HYDRATE	FIBER	ASH	REF.
SAP									
Sap, Maple	252	(348)	33.0	-	-	65	-	0.7	1
GREENS									
Amaranth	36		86.9	3.5	0.5	6.5	1.3	2.6	1
Dock	28		90.9	2.1	0.3	5.6	0.8	1.1	1
Lambsquarters	43		84.3	4.2	0.8	7.3	2.1	3.4	1
Mushrooms (commercial)	35		89.1	1.9	0.6	6.5	1.1	1.0	1
Onions, green	36		89.4	1.5	0.2	8.2	(1.2)	0.1	1
Pokeberry shoots	23		91.6	2.6	0.4	3.7	-	1.7	1
Purslane	21		92.5	1.7	0.4	3.8	0.9	1.6	1
Swamp cabbage	29		89.7	3.0	0.3	5.4	1.1	1.6	1
FRUIT									
Apple	58	(355)	84.5	0.2	0.6	14.5	0.6	0.3	1
Blackberries	58	(355)	84.5	1.2	0.9	12.9	4.1	0.5	1
Blueberries	62	(351)	83.2	0.7	0.5	15.3	1.5	0.3	1
Cherries (sour)	58	(338)	83.7	1.2	0.3	14.3	0.2	0.5	1
Crabapples	68	(342)	81.1	0.4	0.3	17.8	0.6	0.4	1
Cranberries	46	(361)	87.9	0.4	0.7	10.8	1.4	0.2	1
Currants	50	(332)	85.7	1.4	0.2	12.1	3.4	0.6	1
Elderberries	72	(339)	79.8	2.6	(0.5)	16.4	7.0	0.7	1
Gooseberries	39	(334)	88.9	0.8	0.2	9.7	1.9	0.4	1
Grapes	69	(356)	81.6	1.3	1.0	15.7	0.6	0.4	1
Haws, scarlet	87	(341)	75.8	2.0	0.7	20.8	2.1	0.8	1
Pawpaws	85	(345)	76.6	5.2	0.9	16.8	-	0.5	1
Persimmon	127	(338)	64.4	0.8	0.4	33.5	1.5	0.9	1
Pumpkin	26	(294)	91.6	1.0	0.1	6.5	1.1	0.8	1
Raspberries	73	(361)	80.8	1.5	1.4	15.7	5.1	0.6	1
Squash, summer	19		94.0	1.1	0.1	4.2	0.6	0.6	1
Squash, winter	50	(318)	85.1	1.4	0.3	12.4	1.4	0.8	1
SEEDS									
Corn, field	348		13.8	8.9	3.9	72.2	2.0	1.2	1
Lambsquarters	320		13.4	13.3	5.6	45.9	14.6	7.2	2
Maygrass	370		10.7	5.3	6.4	54.3	3.0	-	3
Pumpkin	553		4.4	29.0	46.7	15.0	1.9	4.9	1
Squash	560		4.8	24.0	47.3	19.9	3.8	4.0	2
Sumpweed	535		5.1	32.3	44.5	11.0	1.5	5.8	2
Wild rice	353		8.5	14.1	0.7	75.3	1.0	1.4	1
NUTS									
Beechnut	568		6.6	19.4	50.0	20.3	3.7	3.7	1
Black walnut	621		2.9	24.1	58.5	10.8	1.0	2.7	2
Butternut	629		3.8	23.7	61.2	8.4	-	2.9	1
Hazelnut	634		5.8	12.6	62.4	16.7	3.0	2.5	1
Hickorynut	673		3.3	13.2	68.7	12.8	1.9	2.0	1
Hickorynut, shagbark	692		2.2	11.0	72.7	10.6	1.5	2.0	2
Oak, white	221		47.3	2.8	3.3	43.9	1.3	1.4	2
Oak, red	299		38.2	3.4	12.9	42.1	1.9	1.5	2
Pecan	687		3.4	9.2	71.2	14.6	2.3	1.6	1
ROOTS AND TUBERS									
Duck potato	123	(353)	66.9	4.4	0.8	24.9	1.0	2.0	2
Groundnut	109	(353)	70.7	4.1	1.0	18.6	3.5	2.1	2
Cattail roots (dried)	-	367	7.6	6.9	3.1	79.8	NA	2.6	2
Jerusalem artichokes	7-75		79.8	2.3	0.1	16.7	0.8	1.1	1

Source: 1=Watt and Merrill 1963, 2=Asch and Asch 1978, 3=Crites and Terry 1984).

*Food value dried, parentheses indicate estimation based on published food value and water content.

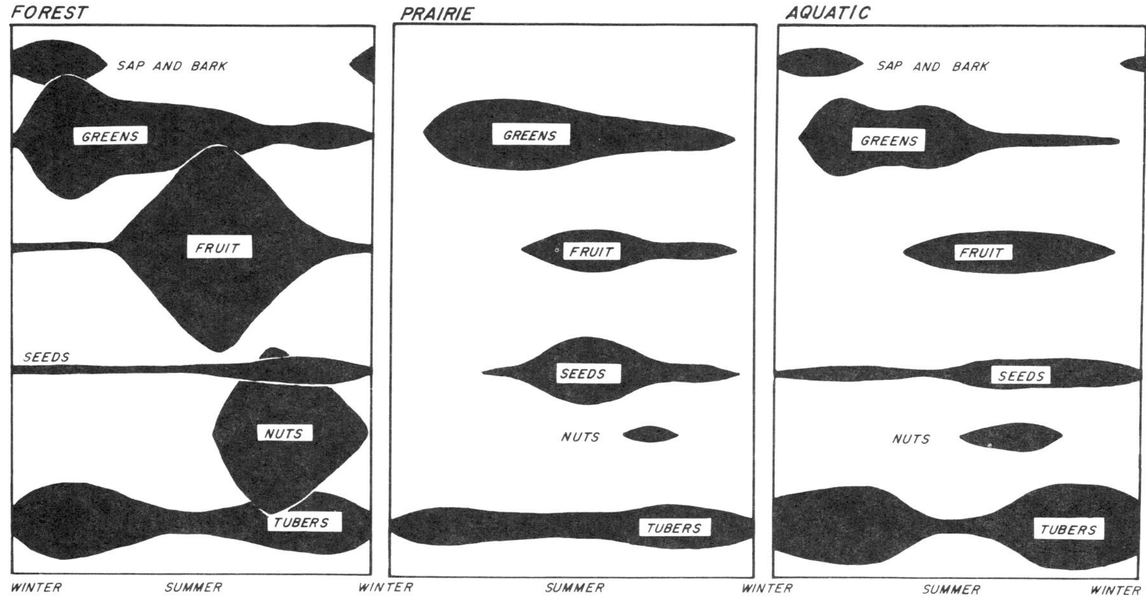

16. Seasonal availability of various potential plant food resources in forest, prairie, and aquatic environments.

The term "fruit" refers to the structure into which the ovary and associated parts develop after fertilization; the fertilized ovule becomes the "seed" (Esau 1960:314). There are numerous kinds of fruit which can be divided into dry and fleshy types. These are further divided based on various anatomical characteristics. Fleshy fruits include the berry (tomato, grape), the drupe (plum, cherry), pome (apple), pepo (squash), accessory (strawberry), and aggregate (raspberry) types.

Dry fruits include those that split at maturity, such as legume (pea, bean), follicle (milkweed), and capsule (poppy), and those that do not split, such as nuts (acorn, hazelnut), achene (sunflower), caryopsis (grass), and samara (maple). Thus the term "berry" is technically inaccurate when applied to blackberries or strawberries and the term "seed" is inaccurate when applied to sumpweed, sunflowers, or grass unless we are referring to the ovule itself with the enclosing parts removed.

While the terms commonly used to describe plant foods may not be technically correct, most are descriptive and generally adequate. For example, most fleshy fruits have approximately the same nutritional content (Table 2); so there is no real reason to differentiate them further. Likewise, many of the small dry fruits we label as "seeds" when we are discussing aboriginal plant usage also have approximately the same food value (depending on whether they are high in starches or oils) regardless of whether they are seeds, achenes or caryopses. Thus, plant foods can be classified into several general categories: sap and cambium; bulbs, roots and tubers; greens or potherbs; flowers, flower buds and pollen; fleshy fruit and "berries"; "seeds" and nuts. Nonvascular economic plants include lichens and fungi, both of which include edible species.

Because the members of each class of plant food have the same function for the plant, they are generally available at about the same time of the year (Fig. 16). Sap flows most heavily and cambium (inner bark) is most edible in the early spring when the tree comes out of its winter dormancy. For several weeks sap can be collected as it flows upward from the roots to initiate flowering and leaf production.

"Greens" or "potherbs" are general terms for the young leaves or shoots of plants used for food. The majority of such plants are most edible (i.e., mild flavored and tender) in the spring when growth is beginning. Even modern garden greens such as lettuce become bitter later in the season.

Fleshy fruits are predominately the product of trees and shrubs, many ripening in the early summer, others during the late summer or fall. Early maturing fruits often have seeds which germinate readily and produce new plants during the same growing season

TABLE 3. Number of Potential Food Plants by Community

COMMUNITY	NO. OF PLANTS	RANK	REFERENCE
FOREST (TABLE 10)			
Dry Upland Forest			
1. Southern Ill. Ridgetops	31	11	Mohlenbrock 1967
2. Driftless Area	14	33	Wunderlin 1966
Dry-Mesic Upland Forest			
3. Driftless Area	13	37	Wunderlin 1966
4. Southern Ill.	49	2	Mohlenbrock 1967
5. Shawnee Hills	30	13	Heineke 1978
Mesic Upland Forest			
6. Prairie Groves	41	5	Calef 1953
7. Eastern Ill.	34	9	Hudnut 1952
8. Eastern Ill.	23	21	Phillippe & Ebinger 1977
9. Ravine, Shawnee Hills	43	4	Heineke 1978
10. Southern Wisc.	31	11	Curtis 1959
11. East-Central Ill.	24	18	Phillippe & Ebinger 1977
Streamside Forest			
12. Central Ill.	38	7	Calef 1953
13. Southern Ill.	13	37	Mohlenbrock 1959
Floodplain Forest			
14. East-Central Ill.	14	33	Crites & Ebinger 1969
15. Southwest Ill.	38	7	Mohlenbrock 1967, et al. 1961
16. Southern Ill.	13	37	Mohlenbrock et al. 1961
17. Central Ill.	27	15	A. Jones & Bell 1974
18. Southern Wisc.	24	18	Curtis 1959
Upland Flatwoods			
19. Southern Ill.	30	13	Voigt & Mohlenbrock 1964
PRAIRIE AND SAVANNA (TABLE 11)			
20. Dry Prairie	3	61	Sampson 1921
21. Mesic Prairie	15	31	Sampson 1921
			Gleason 1910
			A. Jones & Bell 1974
22. Wet Prairie	10	47	Sampson 1921
23. Miss. River Floodplain	6	54	Turner 1936
24. Ill. River Floodplain	6	54	Turner 1936
Sand Prairie			
25. Dry Sand Prairie	12	40	Gleason 1910
26. Wet Sand Prairie	5	57	Gleason 1910
27. Wet Prairie, Lake Michigan	11	44	Gates 1912
Hill Prairie			
28. Loess Hill Prairie	18	27	Evers 1955
29. Glacial Drift Hill Prairie	10	47	Ebinger 1981
30. Gravel Hill Prairie	25	17	Fell & Fell 1956
Savanna			
31. Savanna	14	33	Madany 1981
32. Dry Sand Savanna	14	33	Madany 1981
33. Black Oak Savanna	39	6	Gleason 1910
34. Sand Thicket	16	29	Gleason 1910
35. Bur Oak	20	24	Gleason 1910
36. Mixed Oak	21	22	Gleason 1910
37. Barrens	8	50	Madany 1981
AQUATIC (TABLE 12)			
38. Marsh, Western Ill.	15	31	Solomon 1979
			Sampson 1921
39. Marsh, Central Ill.	21	22	A. Jones & Bell 1974
40. Swamp, Southern Ill.	16	29	Mohlenbrock et al. 1974
			Mohlenbrock 1959
41. Cypress Swamp, Southern Ill.	3	61	Anderson & White 1970
42. Bogs	19	25	Sheviak & Haney 1973
43. Fens	19	25	Curtis 1959
44. Sedge Meadow	18	27	Sherff 1913
			Curtis 1959
45. Seeps and Springs	5	57	Voigt & Mohlenbrock 1964
46. Hillside Marsh, Eastern Ill.	11	44	Phillippe & Ebinger 1977
			Phipps & Speer 1956
47. Floodplain Ponds	12	40	Wunderlin 1966
48. Sinkholes, Southern Ill.	6	54	Voigt & Mohlenbrock 1964
49. Oxbow Lake	5	57	Weik & Baker 1975
50. Prairie Potholes	27	15	Glenn-Lewin & Crist 1981
51. Ponds	3	61	Vogel & Ebinger 1979
52. Sangamon Drainage	7	52	A. Jones & Bell 1974
53. Lake, Southern Wisc.	8	50	Curtis 1959
54. Lake, Southern Ill.	12	40	Mohlenbrock et al. 1961
55. Lake, Eastern Ill.	7	52	Phillippe & Ebinger 1977
PRIMARY AND DISTURBANCE (TABLE 13)			
56. Sandstone Ledge	9	49	Winterringer & Vestal 1956
57. Limestone Glade	81	1	Ozment 1967
			Evers 1955
			Kurz 1981
58. Upper Beach, Lake Mich.	12	40	Gates 1912
59. Disturbance, Central Ill.	11	44	A. Jones & Bell 1974
60. Disturbance, Southern Ill.	48	3	Mohlenbrock 1967
			Heineke 1978
61. Disturbance, Eastern Ill.	34	9	Phillippe & Ebinger 1977
62. Central Miss. Sand & Mudflats	24	18	Evans 1979
63. Disturbance, Southern Wisc.	5	57	Curtis 1959

NOTE: See tables 10-13 for specific plant foods for each community.

while those produced later in the year often must overwinter and germinate the following spring.

Although produced also by many trees and shrubs, small dry seeds are typical of annual or perennial herbaceous species. Annual plants must complete their life cycles in a single growing season. Because of the amount of growth that must be accomplished prior to flowering or fruiting, seed production is often delayed until late summer or fall compared to early spring flowering of many woody plants.

Most nuts (acorns of the white oak group being one exception) overwinter and germinate during the following spring when conditions are optimum. The large nutmeat helps to insure that many nuts will be collected and stored by animals such as squirrels. The thick, hard ovary wall increases the likelihood that some will escape predation. Those that are not eaten begin growth in the spring, often with the advantage of having been buried in the soil by an animal.

Bulbs, tubers, roots, corms, and rhizomes are underground portions of the plant. Some of these parts store food produced by the above-ground leaves during photosynthesis while others are used for the intake of water and minerals. Such plant storage structures are fullest between the fall, just after the completion of annual growth, and the spring, prior to the resumption of growth. Some plants become dormant during the summer so that they are conspicuous (and therefore available unless their precise location has been previously marked) only during the spring and fall.

The reproductive structures, the seeds, nuts, acorns, and fruit, are among the most important plant foods. As already mentioned, fruits and berries have evolved dispersal mechanisms which include producing juicy and sweet berries. Although such fruits generally contain a high percentage of water, they do add considerable variety to the diet and many can be dried and stored for winter use. As far as human use is concerned, the seed either generally goes unnoticed (strawberry) or is discarded (plum, cherry). Even before intentional cultivation of plants, many edible species probably occurred around aboriginal encampments where their seeds had been discarded.

Seeds contain the essential nutrients necessary to enable the newly germinated seedling to become established. For this reason seeds are high in fats and starches. Fat in particular contains much more energy than is available from greens or fleshy fruit. Therefore, seeds, whether they are small such as sunflower or knotweed or large such as hickory nuts, are potentially very important food resources.

NUMBER OF ECONOMIC PLANTS

Portions of Illinois were originally covered by prairie and by beech-maple, oak-hickory, western mesophytic, and southern forest. In contrast, the Great Lakes Region discussed by Yarnell (1964) consists primarily of oak-hickory, beech-maple, and coniferous forests (Fig. 1; Yarnell 1964:4). The number of vegetation types is greater in Illinois than in the more northern Great Lakes Region and the number of plant species correspondingly higher.

Mohlenbrock (1975) lists 2,699 plant species for Illinois, of which 1,916 are native (Ebinger and Lehnen 1981); 29% of the modern flora of Illinois is introduced! Yarnell (1964) estimated that 1,973 native plants occur in the Great Lakes Region, of which 20% (373) are mentioned in the ethnobotanical literature. Of the native plants of Illinois, 24% (465) are mentioned in the ethnobotanical literature pertaining to the Midwest. It is interesting that although the number of economic species is greater in Illinois than in the Great Lakes region, the percentage of plants used aboriginally remains nearly the same, possibly reflecting the percentage of plants which have an inherent economic value.

GEOGRAPHIC DISTRIBUTION OF PLANT RESOURCES

Some habitats have a much greater variety of potential food plants than do others (Table 3). In general, the more extreme the environment (wet or dry), the smaller the number of species, including food plants, that can tolerate it. Mesic communities tend to have more species because they include tolerant dry-adapted and wet-adapted species as well as mesophytic ones. Forests tend to have more potential food plant species than prairies or aquatic communities. In a ranking of the 63 plant communities discussed in the Appendix based on the number of potential food plants (Table 3), no aquatic community ranks higher than fifteenth and no prairie community higher than seventeenth.

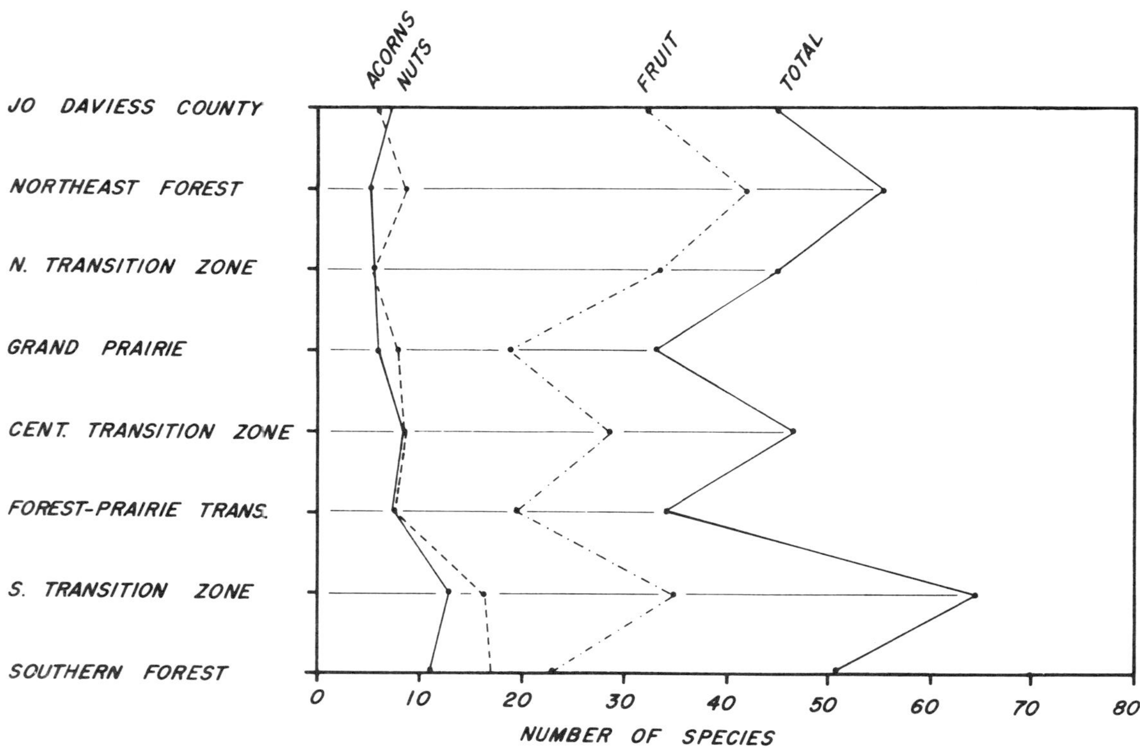

17. A comparison of the number of acorn-, nut- and fruit-producing species for the floristic provinces of Illinois (based on a sample of 95 species).

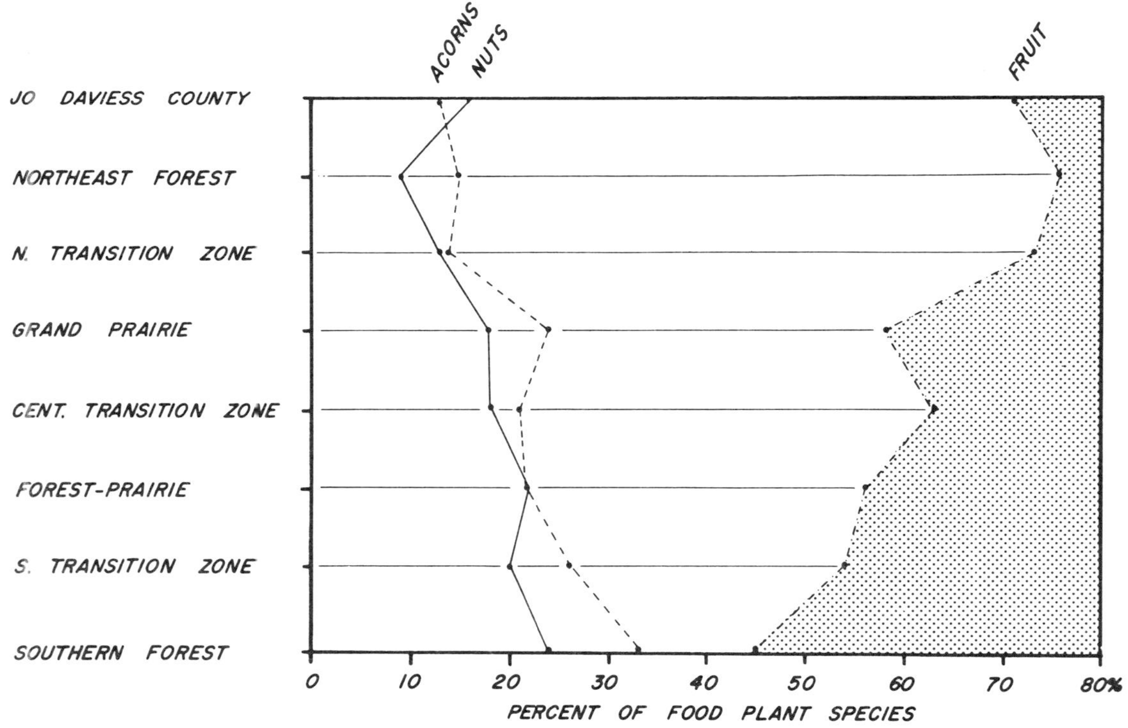

18. A comparison of the percentages of food plant species comprised of acorn-, nut-, or fruit-producers for the floristic provinces of Illinois.

In addition to the relative lack of food resources and the desiccating effects of sun and wind, prairie habitats generally lack water resources and protection from prairie fires, predators, and human enemies. However, prairies do contain many unique plant species that do not occur in forest or aquatic environments and which might serve to draw prehistoric peoples.

Aquatic environments themselves are obviously too wet to normally be used for habitation. In addition, many aquatic communities are breeding areas for mosquitoes and other noxious insects and animals. Aquatic environments have few plant food resources when compared to the forest, ranking no higher than fifteenth in the comparison of total potential food species of various communities (Table 3). However, many plant foods are as unique to wet environments as others are to the dry environments of the prairies. Several plant foods that occur in aquatic habitats were important to aboriginal groups over wide geographical areas (for example, American lotus, cattail, arrowhead).

Aquatic communities vary considerably in the number of potential plant food species they contain. Small bodies of water (seeps, springs, sinkholes, potholes) and those that move rapidly and preclude growth of vegetation have relatively few species. Lakes, ponds, swamps, and marshes generally have a greater number of potential plant food resources, especially during the winter and spring. The collection of aquatic plant foods could be made in conjunction with hunting of migratory waterfowl or fishing in the spring or fall. Historically, many aquatic plant foods were collected throughout the year.

DIFFERENCES IN POTENTIAL FOOD RESOURCES BETWEEN THE FLORISTIC PROVINCES

The potential food plants vary not only between communities but also between geographical regions of the state. The presence or absence of 95 food-producing species was noted for a series of Illinois counties chosen to fall within each of the floristic provinces and transition zones described by Anderson and Ugent (1980). An attempt was made to avoid the counties with major cities or those in which the flora has been intensively studied. To minimize possible error due to greater or lesser numbers of plant collections in various areas, the results for two counties have been averaged for the Northeast Forest, Southern Forest, and North, Central and Southern transitional zones. The number and percentage of nut-, acorn-, and fruit-producing species as well as the total number of species was calculated for each county or pair of counties (Figs. 17 and 18).

Even in such a sampling of only 95 food-producing species, the greater number of species in transitional zones is generally apparent (Fig. 17). An exception is the Northeast Forest Province, represented by Lake County, which has a higher number of species than the Northern Transition Zone. This at least partially reflects the flora associated with the shores of Lake Michigan and the varied glacial terrain of Lake County.

However, the Central Transition Zone counties, with 45 out of 95 species of potential food plants (Fig. 17), average 39% more species than the adjacent Grand Prairie (33 species) and 34.5% more species than the Forest-Prairie Province (34 species). The Southern Transition Zone (64 species) has 87% more species than the Forest-Prairie Province and 26.5% more species than the Southern Forest Province (51 species). It is apparent from this sampling of possible food plants that the diversity of potential food resources in the transition zones between floristic provinces is considerably greater than within individual provinces.

The percentages of nut-, acorn-, and fruit-producing species are also shown for the floristic provinces of Illinois (Fig. 18). In northern Illinois, 71 to 75% of the species were fruit-producing compared to 62 to 54% in the Grand Prairie, Central Transition Zone, and Forest-Prairie Province and 45% in the Southern Forest Province. Nut-producing species are least diverse in the Northeast Forest Province (9%) and become more numerous in southern counties, ranging from 18 to 24% of the food plant species tabulated. Oaks show a similar pattern, a relatively low diversity of species in the Northeast Forest Province (15.4%) and increasing to 32% of the food species in the Southern Forest Province.

Table 4 shows the index of similarity values for the counties studied in each of the floristic provinces. The index of similarity is a correlation coefficient used for comparing two samples based on a number of different attributes (in this case the presence or absence of 95 plant species). Index of

TABLE 4. Indices of Similarity Between Nine Counties Representing the Floristic Provinces of Illinois

PROVINCE	NO.	1	2	3	4	5	6	7	8	9
NORTHERN FOREST	1	-	60	61	72	66	56	47	59	45
NORTHERN TRANSITION	2		-	65	64	63	57	55	53	47
GRAND PRAIRIE	3			-	69	72	78	62	58	58
NW GRAND PRAIRIE	4				-	73	71	54	63	48
CENTRAL TRANSITION	5					-	81	73	78	67
W. FOREST-PRAIRIE	6						-	80	67	69
S. FOREST-PRAIRIE	7							-	64	69
SOUTHERN TRANSITION	8								-	80
SOUTHERN FOREST	9									-

Note: Table based on the presence or absence of 95 potential food plants. The higher the index value, the greater the similarity between different provinces (maximum =100, minimum=0).

similarity is expressed as 2w/(a+b) where "a" is the number of species in one county, "b" the number in a second county, and "w" the number occurring in both. The index is frequently multiplied by 100 with values then ranging from 0 when there is no similarity to 100 when there is complete similarity (Curtis 1959:83).

Because the plants present in a county reflect all of the plant communities of that county, the index of similarity values for counties are higher than those for comparisons between separate plant communities. Curtis (1959:83) found that the two most closely related communities of Wisconsin had an index of 78.9 (dry and dry-mesic prairie) and the two least similar (open bog and dry prairie) had an index of only 0.4. For the Illinois counties, the indices of similarity, based on 95 food plants, range from 81 (Central Transition Zone and Western Forest-Prairie Province) to 45 (Northeast and Southern Forest Provinces). In general, counties in floristic provinces have greater indices of similarity when compared with adjacent transition zones than with other provinces. This is especially apparent in the Central Transition Zone where the presence of both northern and southern species results in a higher index of similarity with several provinces.

FOREST PRODUCTIVITY

Trees must attain a certain minimum size before they initiate flowering and fruit production. For example, Figure 19 shows the average diameter at which optimum acorn production starts plotted as a function of abundance and size of white oak trees in a typical undisturbed old-growth forest. As in all old-growth forest situations, the greatest number of trees occurs in the smallest size class. This happens because, although many seedlings emerge each year, mortality is extremely high among young trees and few survive to reach the largest diameter classes. White oaks generally achieve optimum acorn production between 50 and 200 years of age (Fowells 1965) while averaging 0.15 inches (3.8 mm) of growth per year (Johnson 1974). Thus, a tree has to be about 8 inches (20 cm) in diameter before it begins optimum acorn production. Because a large percentage of the trees in a typical forest stand are below the size necessary to produce acorns, basal area (calculated from the diameter four feet above the ground) is a much better indicator of food production potential than is the number of trees. Basal area for tree stands is usually presented as square meters per hectare (m^2/ha).

For most forests in Illinois, the average basal area falls between approximately 20 and 30 m^2/ha (Table 5). Considerably higher basal areas can be supported by certain floodplain forests, notably cypress swamps and those bottomland forests dominated by fast growing-species such as silver maple, cottonwood, and elm. Other forest types, especially those growing on poor soil where moisture or nutrients are limiting factors, have lower average basal areas of between 14 and 19 m^2/ha.

Great variation occurs in the relative basal areas of various types of food-producing tree taxa in different habitats. Figure 20 shows the relative dominance (relative basal area) of sap-, nut-, and acorn-producing tree species at various elevations above the Sangamon River at Allerton Park, Piatt County, in central Illinois. Sap-producing, flood-tolerant silver maple and boxelder trees are responsible for

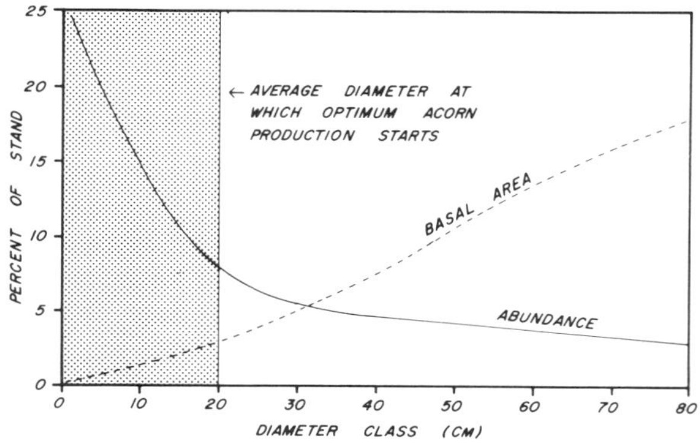

19. Relationship between average diameter at which optimum production of acorns begins for the white oak and relative frequency (based on data from Johnson 1974 and Fowells 1965).

90% of the basal area in the lower, more frequently flooded portion of the floodplain. With increasing elevation, silver maple is replaced by more mesic taxa including oaks and hickories and several fruit-producing species. At Allerton Park, the upland forest is dominated by oak and hickory (Bell 1974).

The pattern seen at Allerton Park is duplicated on a larger scale when the basal areas of trees in various forest types throughout Illinois are compared (Fig. 21). The wettest floodplains have extremely high amounts of silver maple while the driest sites are dominated by oaks. In central Illinois uplands, post oak and scrub oak forests are dominated by oak with basal areas of approximately 10 m^2/ha and greater. Floodplains along the Embarras River in eastern Illinois are dominated by sap-producing species, and have few nut, fruit or acorn producers. Likewise, prairie groves have a large amount of sugar maple along with the oaks and some fruit- and nut-producing species. However, it is the relatively mesic terraces, slopes, and ravines which have the greatest diversity of food-producing tree species and the most balance between sap, fruit, nut, and acorn producers.

The basal area of each type of potential food plant is a good indicator of that area's food-production potential. As for arboreal food resources, wet floodplains are relatively barren after the early spring availability of sap.

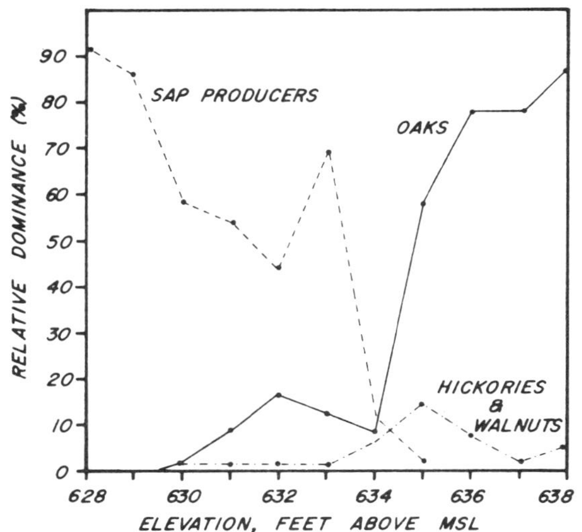

20. **Relative dominance of sap-, acorn- and nut-producing species in relation to elevation of an upland streamside forest at Allerton Park in East-Central Illinois (based on Bell 1974).**

TABLE 5. Basal Area (m2/ha) for Various Forest Types in Illinois

COMMUNITY	AVERAGE	RANGE	NO.	REFERENCES
UPLAND XERIC FOREST	18.7	17.1-21.4	3	Telford 1926 Adams and Anderson 1980
UPLAND MESIC & DRY MESIC	24.5	19.8-33.9	12	Ashby 1968 Telford 1926 McClain & Ebinger 1966 Ebinger 1968 Hughes & Ebinger 1973 Boggess & Geis 1967 Root, Geis & Bogess 1971 Adams & Anderson 1980 Ebinger & Parker 1969
PRAIRIE GROVES	24.9	21.4-27.21	3	Boggess 1964 Boggess & Bailey 1964 Boggess & Geis 1966
UPLAND FOREST-PRAIRIE TRANSITION	25.2	-	1	Jackson & Petty 1971
UPLAND OLD-GROWTH	30.0	-	1	Weaver & Ashby 1971
MESIC BOTTOMLAND	26.1	13.4-43.9	10	Telford 1926 Hosner & Minckler 1963 Lindsey _et al_.1961 Hughes & Ebinger 1973 Adams & Anderson 1980 Nyboer and Ebinger 1976
WET BOTTOMLANDS	43.3	14.9-68	11	Lindsey _et al_.1961 Hosner & Minckler 1963 Telford 1926 Crites & Ebinger 1969
SWAMP	90.16	-	1	Lindsey _et al_.1961
FLOODPLAIN OLD-GROWTH	36.8	-	1	Robertson _et al_.1978

Likewise, dry oak forests produce few arboreal foods other than acorns. The season of exploitation of various habitats and the specific plants used aboriginally must have reflected the forest composition.

Most tree species produce the greatest amount of fruit or nuts when they do not have to compete for light, water, or nutrients. The denser the forest, the greater the competition. Thus, for example, pin oak will produce two to three times as many acorns when the basal area of the forest is 11 m^2/ha as when it is 20 m^2/ha (Fowells 1965:605). Likewise, understory grass and herb production in Missouri pine stands increased four to five times when the stands were thinned from 30 m^2/ha to 16 m^2/ha and 14 times when thinned to 11 m^2/ha (Ehrenreich 1960).

If factors such as soil fertility and moisture availability are held constant, both open-grown trees and their understory species are more productive than those growing in dense forest situations. This being the case, forests with artificially low basal areas, such as might result from periodic fires or thinning, would produce more food for both man and game animals. Naturally thin areas transitional between forest and prairie, such as savanna or barrens, may have had a particularly rich diversity of plant resources.

ENVIRONMENTAL RELIABILITY

The optimum environment for plant

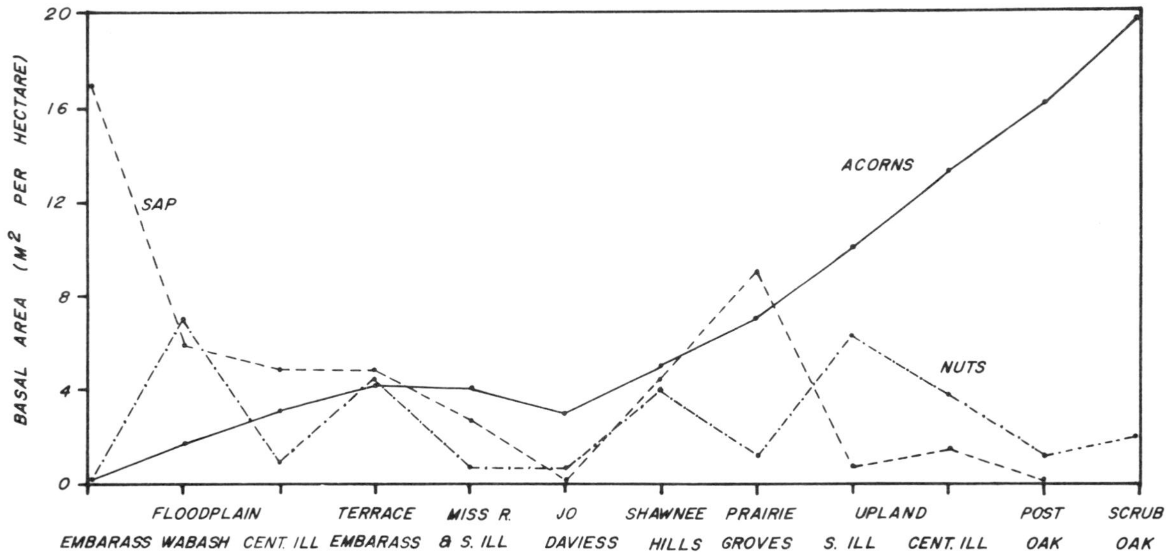

21. **Basal areas of sap-, nut-, acorn-, and fruit-producing species in forest stands throughout Illinois.**

growth depends on the plants under consideration. It is that situation in which any given plant species is capable of the greatest biomass production, including formation of reproductive propagules (fruit, nuts, seeds). The extent to which plants can be relied upon for biomass production (food resources) depends on climate, soils, and plant adaptation to a specific environment. Because of lower amounts of stress (heat, water, nutrient), certain environments are more reliable than others at producing food.

For any large area, environmental reliability is tied to the stability of the climate and/or to the presence of sheltered microenvironments such as those produced by the buffering effect of topographic relief or the presence of large bodies of water. No matter what the landscape, if the environment is stable then the best-adapted plant species will ultimately dominate the communities that evolve. If the dominant species are nut and fruit producers, as in the oak-hickory forest, then a large potential food resource base exists. If nut and fruit producers are not common, as in conifer forest, then the resource base will be considerably smaller. Although not a natural community, one with the maximum food-production potential is an orchard comprised of thousands of identical trees. The high ratio of fruit production to energy expended in orchards makes them extremely profitable ventures, especially when compared to the alternative of going through a forest collecting fruit from scattered trees.

If the climate is variable, however, or if there is a sudden sharp change from one climatic regime to another, the effects of topographic relief can become exceedingly important. Varied landforms tend to create many diverse microenvironments and to modulate the strength of climatic impacts on those environments. If there is adequate topographic diversity, then the majority of climatic extremes are buffered in at least some situations. Although overall production of specific food resources in a diverse environment might not be so great under favorable conditions as it would in an area of higher uniformity, the resources remaining available during periods of climatic stress would probably be greater. To continue with the orchard analogy, periodically most of the Florida citrus crop is destroyed by freezing temperatures or the Illinois apple or peach crop is destroyed by late spring frosts. Even in an orchard, however, a few particularly well-adapted trees in more sheltered locations will survive such extremes. In most forest situations, the diversity of environments is much greater than in an orchard, meaning that the chances of any individual tree's being unaffected by extremes are also greater. In addition, of course, environmental variability creates resource diversity and a larger resource base. Even though only a small number of plant resources might have been used aboriginally, the diversity of plants would also be reflected in a greater diversity and abundance of animals.

Of all environments, aquatic and semiaquatic habitats are unique in their degree of independence from minor climatic fluctuations. The most important adaptation of aquatic plants is the adaptation to constant or long-term submergence, which would kill the majority of less well adapted plants. Because water changes temperature slowly, it will remain cold later in the spring and warm longer in the fall than adjacent land surfaces and will have little daily fluctuation. Aquatic plants, therefore, are never so hot nor so cold as those on dry ground. Likewise, drought affects aquatic environments only under the most severe conditions when the water sources themselves disappear. Aquatic plants tend to be perennial and to die back to the roots during the winter. They often have large fleshy storage roots or tubers that are available for collection as food throughout the year. The mean net primary productivity of swamp and marsh is 2000 $g/m^2/yr$ while that of lakes and streams averages 250 $g/m^2/yr$ compared to 1200 $g/m^2/yr$ for temperate deciduous forest or 600 $g/m^2/yr$ for grassland (Table 6, Whittaker 1975:Table 5.2).

The floodplains of larger streams and rivers are another type of semiaquatic environment that is resource rich and relatively reliable. In the annually inundated floodplains, the majority of the annual moisture is received by flooding during spring runoff. Because drainage basins collect and funnel water, lower parts of the basin are virtually certain of obtaining moisture even in very dry years, although the duration of flooding may vary. In addition, flood-deposited materials tend to be rich in nutrients. Throughout history, it has generally been the floodplains where human settlement and plant cultivation first occurred.

Although certain trees, such as silver maple, can withstand prolonged flooding, the dominant plants on the newly deposited muds of seasonally flooded areas are annual species capable of completing their life cycles even if additional moisture is not received. With some exceptions, these plants tend to be weedy types capable of rapidly invading and inhabiting newly disturbed habitats. Because they are "pioneer" species and adapted to marginal environments, their seed production is tremendous. The seeds of aquatic, marsh, and mudflat plants are primary food sources for great numbers of migratory waterfowl.

Because of the geologic and glacial history of Illinois, the primary sources of permanent water outside the limit of Wisconsinan glaciation are the rivers and streams. Lakes and ponds are limited almost entirely to the broad floodplains of the major stream valleys; water sources such as springs and sinkholes are extremely localized.

TABLE 6. Mean Net Primary Production and Mean Biomass Per Unit Area for Different Ecosystem Types

ECOSYSTEM TYPE	MEAN NET PRIMARY PRODUCTION g/m2/year	MEAN BIOMASS PER UNIT AREA (kg/m2)
Tropical rain forest	2200	45
Tropical seasonal forest	1600	35
Temperate evergreen forest	1300	35
Temperate deciduous forest	1200	30
Boreal forest	800	20
Woodland and shrubland	700	6
Savanna	900	4
Temperate grassland	600	1.6
Tundra and alpine	140	0.6
Desert and semidesert scrub	90	0.7
Extreme desert, rock, sand and ice	3	0.02
Cultivated land	650	1
Swamp and marsh	2000	15
Lake and stream	250	0.02

Source: Whittaker 1975, Table 5.2.

Prairies are perhaps as reliable, if not as productive of potential human food resources, as the aquatic and semiaquatic environments. Prairie plants are well adapted to their environments, and the biomass produced annually on the prairie is capable of supporting large populations of animals. Prior to Euro-American settlement, herds of bison traversed the plains and prairies, primarily to the west of the Mississippi River, taking advantage of the prairie vegetation. The reliability of the prairie as a resource is demonstrated by both the size of the herds and the well-established migration routes that evolved with time.

In general, forest resources are less reliable than those of either aquatic or prairie environments of Illinois. The climate of the Prairie Peninsula is better suited for the growth of grasses and herbs, and forest is frequently limited to broken and thus sheltered topography. Many trees produce abundant fruit and nut crops only every 2 to 3 years and insects can be a serious problem, especially during years of small crops. However, the diversity of plant and animal resources in the various forest communities is large and the quality (food value) high. Mean net primary productivity in temperate deciduous forests is 1200 $g/m^2/yr$ (Table 6, Whittaker 1975: table 5.2). However, of that only about 2% is in the form of fruit and flowers (Whittaker 1975: table 5.1).

Once again, topographic relief is important because it creates a variety of habitats. Although much of the state is flat-lying, portions of northern and southern

22. Average slopes for Illinois (after Fehrenbacher *et al*, 1968).

Illinois, as well as areas adjacent to the major streams, show relatively high topographic relief (Fig. 22). The major streams in this region (the Mississippi, Illinois, Wabash) run primarily north to south while secondary streams tend to flow east or west. The valley slopes of these secondary streams face north and south, effectively creating a temperature differential of 7 to 8° F between slopes. An even greater temperature differential can occur between a ravine bottom and the adjacent upland. For example, on a July day at Giant City State Park in Jackson County, a 25° F temperature difference was recorded between the blufftop and the shaded underside of an overhanging ledge (Voigt and Mohlenbrock 1964: 12). Due to sun exposure, south- and west-facing slopes contain relatively xeric plant communities while the more sheltered north- and east-facing slopes support more mesic species. Dry open woodlands, however, generally have more species than mesic closed forests because the denser canopies of the latter suppress the growth of understory plants.

The amount of available moisture is an important factor in plant growth and is determined by precipitation, loss due to runoff and evaporation, and the ability of the soil to retain water. The combination of water removed by evaporation and transpiration is termed *evapotranspiration*. When maximum evapotranspiration is less than precipitation, there is a potential water surplus; when maximum transpiration exceeds precipitation, there is a water deficit.

For normal years, no area of Illinois has a shortage of water for evapotranspiration (Fig. 23); water is not a limiting factor for plant growth. Considerable variability in potential water surplus exists in the state, however, with more than 11 inches in portions of extreme southern Illinois and less than 3 inches in parts of northeastern and west-central Illinois. These areas with low potential water surplus in normal years also tend to show the greatest water deficit in dry years (D. Jones 1966: Fig. 18).

The amount of water available for plant growth is not determined entirely by precipitation and evapotranspiration, however. Topography and the depth and type of soil are also important in determining the amount of water retained for potential plant use. The degree to which an interaction of climate and soil factors affect crop yields is shown in Figure 24 (Changnon *et al*. 1982: Fig. 1). (Soil

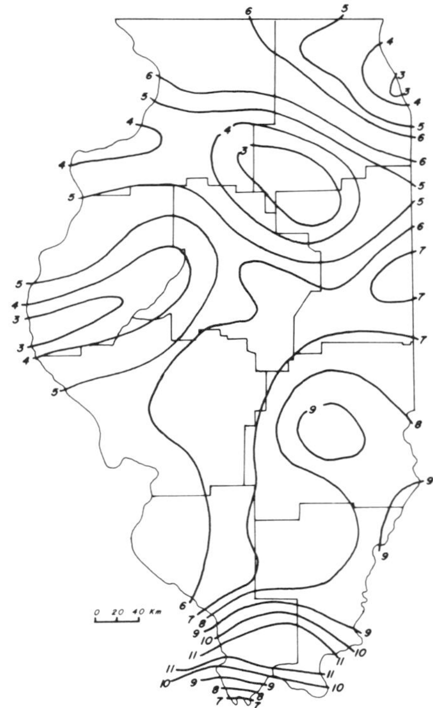

23. Mean annual difference between precipitation and potential evapotranspiration (potential water surplus) in inches (after D. Jones 1966).

factors limiting plant growth are shown in Fig. 13.) Most of southern Illinois, characterized by shallow soils underlain by bedrock or by a claypan or fragipan layer, and portions of central and east-central Illinois, characterized by sandy and gravelly soils, are greatly affected by climatic variations. Portions of west-central Illinois, characterized by relatively good soils but rugged topography and low potential water surplus, are also seriously affected by climatic oscillations (Fig. 24).

The most stable environments for the production of human food resources are those characterized by relatively high potential water surplus and good soils. For many plants, topographic relief also provides protected microenvironments that may act as buffers against extremes of climate.

These environmental factors may have been significant in the development of aboriginal subsistence and settlement patterns in Illinois. Some of the most important archaeological sites in the state occur along the Mississippi and lower Illinois rivers in areas where weather has a relatively great effect on crop yields (Fig. 24). The Lower Illinois Valley is of particular interest because of the

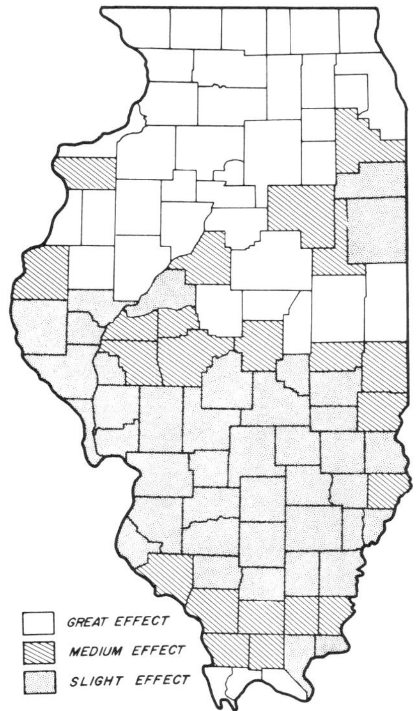

24. Map showing the degree to which weather affects crop yields in Illinois counties (after Changnon *et al.*, 1982).

relatively early appearance of cultigens and the relatively heavy use of native seed-producing plants (Asch *et al.* 1972, Styles 1981). Human adaptation that occurred in the Lower Illinois Valley may, in part, reflect the relatively major effects of climatic variability on a region that was optimum for human settlement. If upland and hillslope resources were severely diminished by climatic extremes, such as those that occurred during the Hypsithermal, the impetus would surely be towards greater utilization of relatively stable floodplain habitats.

SPECIES RELIABILITY

Certain plant foods are reliable merely because of the plant part from which they are derived; roots, leaves, stems, and sap are essential to the plant's survival and must therefore be present each growing season. Other plant structures (seeds, fruit, flowers, nuts) are formed to disseminate the species, and the species will continue to survive if the seeds are not formed for several years. Reliable species are relatively unaffected by insects or disease, and their food products are produced either continually or on a predictable schedule. Roots and tubers are available more or less continuously; sap flows heavily and plants renew growth producing "greens" in the spring; and herbaceous annuals produce seeds every year.

The most reliable food-producing species are those that either have a wide range of ecological tolerances and occur in a wide variety of environments or those that are tied to a narrow but stable environment. It is often the case that the most widespread species are less capable of competing in a specific environment and therefore less abundant than other species which may be extremely competitive over a narrow range of conditions. However, this might not be true during periods of relatively rapid climatic change when the ability to adapt would be at a premium while specialized plants might be displaced as the habitats evolved. The environmental tolerances of most plants were probably severely tested during the mid-Holocene Hypsithermal period.

The response of plants to cold and drought are so similar that plants immune to one are usually immune to the other (Daubenmire 1959:194). For this reason dry winters, a normal condition of the Prairie Peninsula, have relatively little effect on plant distribution compared to dry growing seasons.

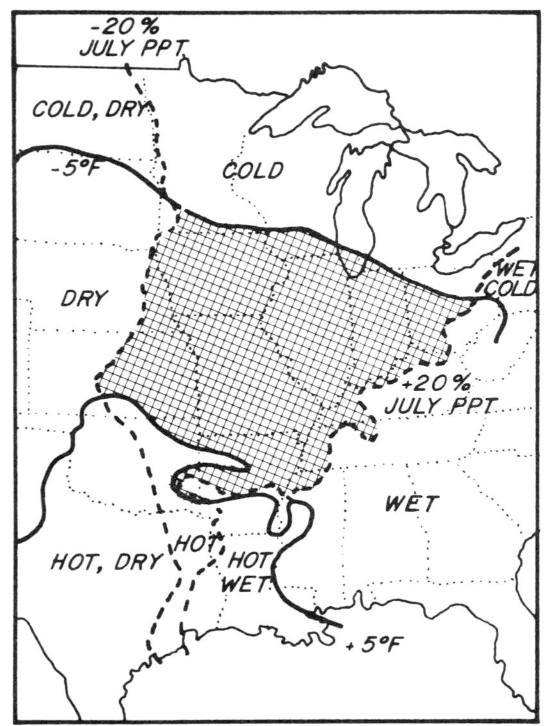

25. Map showing areas of the Midwest where the average July temperature is 5° F higher or lower and the average July precipitation is 20% higher or lower than for Springfield, Illinois.

Summer climate, on the other hand, is especially important to plant distributions in the Prairie Peninsula. Species adapted to the prairie climate, such as the grasses, are often able to grow and reproduce on the basis of a single rainfall. Trees, which normally make their growth on the basis of winter precipitation, are susceptible to the shortfalls in summer rainfall that commonly occur in this region (Transeau 1935). To examine the possible effects of summer climate more closely, Figure 25 shows the areas of the central United States that have July temperatures of 5° F higher or lower and precipitation 20% higher or lower than Springfield, Illinois. These ranges of temperature and precipitation probably approximate maxima that have occurred throughout the Holocene (Wendland and Bryson 1974), and these boundaries probably demarcate the area in which we might seek analogs for the Holocene vegetation of Illinois. Within this range fall all of Iowa and Missouri and most of Indiana as well as portions of surrounding states. It appears that the early Holocene was probably no wetter or colder than western Ohio is today while the Hypsithermal was probably no drier or hotter than eastern Kansas.

The effects of various combinations of these July temperature and precipitation regimes on the distribution of 13 potentially important arboreal food resource species that presently occur in central Illinois are shown in Figure 26. None of these 13 tree species presently occurs under a July climate that is both 20% drier and 5° F hotter (equivalent to portions of central Kansas and Nebraska presently covered by grassland) than modern central Illinois. Since plant remains of most of these species occur in Archaic Period archaeological sites in Illinois, it can be inferred that Hypsithermal climatic conditions were therefore probably not that extreme.

Baldwin Woods, a stand of oak-hickory forest in eastern Kansas (Wells and Morley 1964), is potentially comparable to what existed in central Illinois during the Hypsithermal. Although the climate is relatively hot and dry, Baldwin Woods, on a moderate north-facing slope with deep soils, had 31 species of trees and 21 species of shrubs and vines. Even though, as Braun (1947) notes, the numbers of forest species and their abundances diminish in a westerly direction, this is a greater diversity than occurs in the oak-hickory forest of southern Wisconsin (Wells and Morley 1964, Curtis 1959). It is also an excellent illustration of the effects of sheltered environments.

The relative frequency of potential food plants in Baldwin Woods are: shagbark hickory 25.2%; red oak 4.7%; black walnut 4.7%; and bur oak 3.9%. Other trees with edible fruit or nuts include mulberry, sugar maple, shellbark hickory, persimmon, black cherry, white oak, and black oak. Nut- and fruit-producing shrubs include gooseberry, pawpaw, grape, blackberry, hazelnut, crabapple, and elderberry. Even in eastern Kansas, virtually all of the common nut- and fruit-producing genera of central Illinois continue to occur in select environments.

Baldwin Woods is somewhat unique for so westerly a site, partly because the soils of Kansas tend to be poor. Good soils, on the other hand, are the rule in central Illinois rather than the exception and they were probably much more extensive during the early Holocene. Pollen evidence from central and northern Illinois suggests that mesic forests were fairly common on protected slopes during the early Hypsithermal when upland forests were being replaced by prairie and depauperate oak-hickory forest (J. King 1981). Since many of the same plants

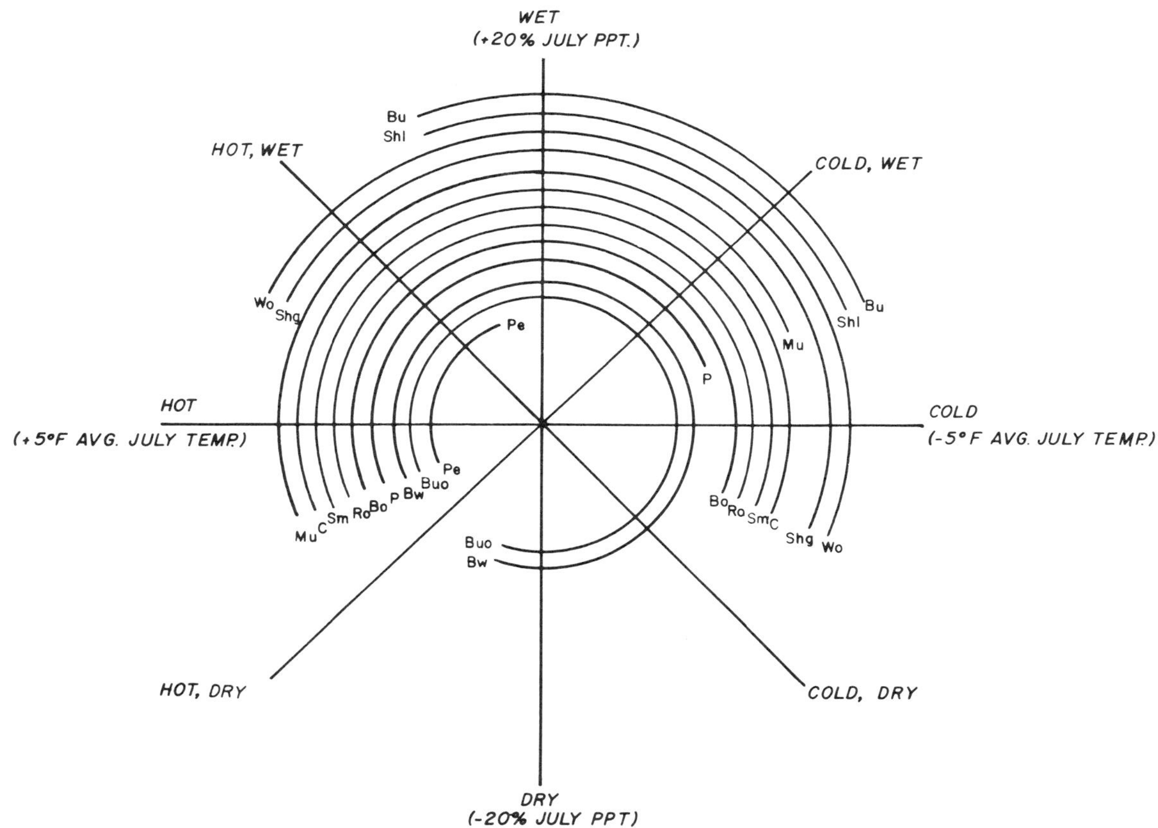

26. Relationship between July climate and tree distribution, based on modern climate and ranges. Pe=pecan, BuO=bur oak, Bw=black walnut, P=persimmon, Bo=black oak, Ro=red oak, Sm=sugar maple, C=black cherry, Mu=mulberry, Shg=shagbark hickory, Wo=white oak, Shi=shellbark hickory, Bu=butternut.

occurring in central Illinois also occur in eastern Kansas under considerably drier conditions, the available plant resources therefore probably have remained much the same, although varying through time in relative abundance.

NATIVE AMERICAN PLANT USAGE

HUMAN SELECTION OF PLANT FOODS

Enough potential food resources exist in most situations that all need not be used and choices can be made among them. While such decisions are being made, there is a selective advantage to those choices which reduce the ratio of energy expended in subsistence pursuits to energy intake, thereby allowing more time for other activities. Because the human pattern of food exploitation in a given habitat is frequently similar to that of other animals (Bourliere and Hadley 1970:4-5), certain parallels can be drawn between ecological and economic approaches to resource selection, and optimal predator strategies have been defined and applied to human subsistence behavior (Rapport and Turner 1977).

The strategy frequently used to describe human behavior is that of attempting maximum energy input and minimum energy expenditure (B. Smith 1975:139). If subsistence-oriented societies follow a maximizing strategy in which human energy requirements are met with a minimum of effort, then at low population densities a group would rely on a limited number of low-cost resources (Earle 1981, Christenson 1981). Additional resources would be added as the need arises, with each additional resource requiring proportionately higher amounts of energy expenditure for the same energy input. This process continues unless/until a resource with a sufficiently high potential yield is encountered to make it more energy efficient to focus on the higher-yielding resource than to continue to exploit a wide diversity (Christenson 1981).

The abundance, diversity, and quality of available food resources vary between seasons and habitats, especially if food is not being stored. A food resource that might be insignificant during some seasons or in some ecological situations might be important in others. For example, during the fall greens and tubers might be ignored in favor of nuts with a relatively high food value. However, both tubers and greens might, in turn, become of value during the late winter and spring after the nuts have been consumed. Likewise, a specific food resource with the same abundance in both the prairie and forest might be considered much more valuable in the prairie where the total number of other food resources is relatively low.

Each plant community has an individual suite of plants which would achieve primary economic importance if a human group were going to subsist entirely in that community. The characteristics such plants would share would be relatively high abundance, relative ease of collection and processing, and relatively high food value compared to other plants of those communities. However, a rather uncommon food resource might still be sought out if, for example, the food value was exceptionally high.

Archaeological evidence, as well as high food value, relatively high abundance, and wide distribution, indicates that nuts were a key food resource in the Midwest for thousands of years. Nutshell, primarily that of hickory but with a greater proportion of other taxa in Early and Middle Archaic components, is the dominant type of plant remains from many archaeological sites.

Differing nut-producing taxa vary in abundance between different plant communities. The relative abundance of various taxa in archaeological sites might, therefore, be due to a number of factors, none of which are mutually exclusive and all of which may have varied in importance from site to site. Such factors include:

(1) A change in the season of occupation from the period of availability of one nut taxon to that of another (Asch and Asch 1976).
(2) Climatic change favoring the growth of certain species over those which had previously grown most successfully.
(3) Change in processing techniques (originally proposed by P. Munson and discussed by Styles *et al.* 1983).
(4) Environmental changes, such as those in floodplain morphology, which might create new habitats or alter the proportions of various taxa in the floodplain forest.
(5) Differential migration of various nut taxa into the Midwest following deglaciation. Hickory, for example, did not complete its migration into the Midwest until 8300 YBP (Davis 1976, J. King 1981).
(6) Changes in food preferences.

Not all habitats are equally conducive to the growth of small, starchy, or oily seeds which occur in "disturbed" sites (knotweed, goosefoot, sumpweed, maygrass, little wild barley). Comparison of plant remains from several early Late Woodland sites in the central and lower Illinois River Valley (Asch and Asch 1981, F. King 1983, Munson *et al.* 1971), suggests that intersite variation occurs not only in nut-producing but in seed-producing taxa as well. While aboriginal groups in the region relied on much the same suite of native cultigens, the importance of specific species probably depended on the relative abundance of the various species in the local natural environment and microenvironmental conditions where the plants were encouraged or cultivated (F. King 1983). The actual number of species on which a group might rely would vary depending on the size of the human population, the relative abundance, ease of gathering, food value of various plant and animal resources, and the season.

The strategies used to maximize energy input would influence which plants and animals were exploited as well. The selection of site location by a prehistoric human group is an important decision. Food resources differ between plant communities, and a site lying near the intersection of several communities would be in a good position to exploit resources in all of them. There would also be the added benefit of the greater abundance and diversity that naturally occurs along ecotones, such as the prairie-forest margin (F. King and Graham 1981).

Scheduling of food procurement activities is another important way in which humans and animals conserve energy in subsistence activities (Styles 1981:16). For example, it would be logical to collect a number of resources on one trip to a distant habitat rather than make several trips.

A third strategy available to all animals is migration to a habitat where more food is available. However, while most animal populations are self-regulating and tend to stay within a size for which there are available food resources, human populations have tended to steadily increase. Thus, the ability to migrate is dependent on there being suitable unoccupied habitats available elsewhere.

An additional way of increasing available energy is to gain control over the food supply such as by herding animals or cultivating crops. This would increase both the yield of food in a given area and the likelihood of its utilization by the controlling population (Reed 1977:15). With the advent of plant and/or animal husbandry, scheduling would have to be reevaluated, as would each plant and animal resource, to conform to a new schedule mandated by cultivation/domestication activities.

Table 7 lists Illinois plants that are known cultigens, semicultigens, or potentially important wild food plants in Illinois, including species which have been found at archaeological sites, regardless of economic importance. Several species which were important in other geographical regions have been excluded because their distributions in Illinois are limited and there is no evidence that they were ever more abundant in the past. Some of these plants are virtually the only plant foods available during a certain season, such as maple sap in early spring. All of those plants noted as "important" are included in the ethnobotanies of many groups indicating a general rather than localized utilization.

Of the "important" plants on this list, nuts and acorns have high food values and can be easily stored, making them extremely valuable. Although fresh fruit contains 75 to 90% water and only 37 to 127 calories/100 g (Watt and Merrill 1963), the caloric value of dried fruit increases to 313 to 368 calories/100 g when dried to 5% moisture content (Table 2). While the caloric values of dried fruits are still only about one-half those of nuts or oily seeds such as sunflower, pumpkin, or sumpweed, the values are similar to those of maple sugar, corn and wild or other starchy seeds and higher than those of acorns which require a great amount of processing. The roots and tubers on this list have moderate caloric values; however, they are both common and available throughout the year, particularly during the winter and early spring when other food resources are scanty.

Based on archaeological and historic evidence as well as plant distribution, the plants listed in Table 7 as "important" include those most critical to aboriginal subsistence in Illinois. Although the level of usage depends on the local availability of each plant resource as well as the relative availability, food value, and ease of collection of others and on the specific human group and time period under consideration, all of these plants were probably used by Native Americans in Illinois.

Table 7 also lists plants that have been found in archaeological sites in the Midwest but which were probably not important food species in Illinois. These plants are either uncommon, have low nutritional value or low palatability, require lengthy processing, or were used by few historic groups.

In addition, Table 7 lists potential food plants with preservable parts that have thus far *not* been found in archaeological sites in

TABLE 7. Important Cultigens, Quasi-cultigens and Potential Wild Food Plants of Illinois

SPECIES	PART	HABITAT	ABUNDANCE	NH*	YBC**	STATUS#
Acer saccharum (sugar maple)	sap	rich woods	common throughout	12	1	IW
Amarathus spp. (amaranth, pigweed)	seeds***+	waste ground	common disturbed	4	-	C, IW
Amelanchier spp. (juneberry)	seeds+	wooded slopes	occ. throughout	5	?	IW,U
Amphicarpa bracteata (hog peanut)	seeds***+	woods & thickets	occ. throughout	10	0	IW
Apios americana (groundnut)	tubers***+	thickets & prairies	common throughout	7	0	IW
Arctostaphylos uva-ursi (bearberry)	seeds***+	sand areas	rare, n. Ill.	2	-	IOR,U
Asimina triloba (pawpaw)	seeds***+	low woods, slopes	common s. Ill.	7	3-4	IW
Carya spp. (hickory, pecan)	nuts***+	woods	occ.-common throughout	9	1-2	IW
Castanea dentata (chestnut)	nuts***+	woods	rare, s. Ill.	0	-	IOR,U
Celtis occidentalis (hackberry)	seeds***+	woods	common throughout	16	-	W
Cercis canadensis (redbud)	seeds***+	woods	common, s. Ill.	?	-	U
Chenopodium spp. (chenopod)	seeds***+	disturbed	common	-	1	C,Q
Citrullus vulgaris (watermelon)	seeds***+	-	-	-	-	C
Cornus stolonifera (red osier)	seeds***+	woods	occ. in n. Ill.	4	-	U
Corylus americana (hazelnut)	nuts***+	thickets, dry woods	common throughout	9	?	IW
Crataegus spp. (hawthorn)	seeds***+	woods	occ.-common throughout	5	?	U
Cucurbita pepo (squash, gourd)	seeds***+	-	-	-	-	C

Table 7. (Continued).

SPECIES	PART	HABITAT	ABUNDANCE	NH	YBC	STATUS
Dentaria laciniata (pepperroot)	tubers***+	woods	common	8	-	IOR,U
Diospyros virginiana (persimmon)	seeds***+	dry woods, clearings	occ.-common cent. & s. Ill	9	2	IW
Elymus canadensis (nodding wild rye)	seeds+	dry prairies	common throughout	9	-	IOR,U
Fagus grandifolia (beech)	nuts***+	woods	occ. s. Ill.	5	-	IOR,U
Fragaria spp. (strawberry)	seeds+	prairies, fields	occ.-common throughout	16	1	IW
Gaultheria procumbens (checkerberry)	seeds***+	sandy woods & bogs	rare, n. Ill.	0	-	U
Gaylussacia baccata (black huckleberry)	seeds+	rocky woods	occ. north & south	0	-	IOR,U
Gleditsia triacanthos (honey locust)	seeds***+	woods	common s. Ill.	12	-	U
Glyceria striata (fowl manna grass)	seeds+	moist ground	common throughout	15	-	IOR,U
Gymnocladus dioica (Kentucky coffee tree)	seeds***+	low woods	occ. throughout	6	-	U
Helianthus annuus (sunflower)	seeds***+	-	-	-	-	C
Helianthus tuberosus (Jerusalem artichoke)	tubers	moist ground	common throughout	2	cont.	Q
Hordeum jubatum (squirrel tail grass)	seeds+	disturbed	occ.-common throughout	5	-	IOR,U
Hordeum pusillum (little wild barley)	seeds***+	disturbed	cmm. south. three-fourths	?	-	Q,IW
Ipomoea pandurata (wild sweet potato)	tubers***+	waste ground	occ. - common southern three-fourths	0	-	IOR,U
Iva annua (sumpweed)	seeds***+	moist ground	occ. throughout	?	-	C,Q
Juglans cinerea & nigra (butternut, black walnut)	nuts***+	rich woods	occ. throughout	3	2-3	IW

Table 7 (Continued).

SPECIES	PART	HABITAT	ABUNDANCE	NH	YBC	STATUS
Lagenaria siceraria (bottle gourd)	seeds***+ rind***+	-	-	-	-	C
Lathyrus palustris (marsh vetchling)	seeds+	moist ground	occ. n. Ill.	7	-	U
Malus ioensis (Iowa crabapple)	seeds***+	woods & thickets	occ.-common	4	?	IW
Mitchella repens (partridge-berry)	seeds+	rocky woods	occ. north & south	0	-	IOR,U
Morus rubra (mulberry)	seeds***+	woods	common throughout	10	?	W
Nelumbo lutea (American lotus)	nuts***+ tubers	lakes & ponds	occ. throughout	5	0	IW
Nicotiana rustica (tobacco)	seeds***+	-	-	-	-	C
Nuphar advena (yellow pond lily)	tubers	lakes & ponds	occ. throughout	5	0	IW
Panicum capillare	seeds+	waste ground	common throughout	10	-	U
Passiflora incarnata (passion flower)	seeds***+	dry soil	occ. s. third	0	-	IW,Q
Petalostemum spp. (prairie clover)	seeds+	prairies	occ. throughout	8	-	U
Phalaris caroliniana (maygrass)	seeds***+	-	-	-	-	C,Q
Phaseolus vulgaris (common bean	seeds***+	-	-	-	-	C
Physalis spp. (ground cherry)	greens seeds+	disturbed	common throughout	11	-	W
Podophyllum peltatum (may apple)	seeds+	open woods	common throughout	10	-	W
Polygonum erectum (knotweed)	seeds***+	disturbed	occ.-cmm throughout	?	-	C,Q
Prunus americana (American plum)	seeds***+	woods & thickets	occ. throughout	6	1-2	IW

Table 7 (Continued).

SPECIES	PART	HABITAT	ABUNDANCE	NH	YBC	STATUS
P. serotina (black cherry)	seeds***+	woods	common throughout	16	1-5	IW
Quercus spp. (white oak)	acorns***+	upland and	common throughout	--	4-10	IW
Rhus spp. (sumac)	seeds***+	woods	common throughout	9	-	W
Ribes spp. (gooseberry, currant)	seeds+	woods	common throughout	9	?	IW
Robinia pseudoacacia (black locust)	seeds***+	woods	originally southeastern Ill.	3	-	U
Rosa spp. (wild rose)	seeds***+	woods	common throughout	?	-	U
Rubus sp. (blackberry, raspberry)	seeds***+	woods, edges thickets	common throughout	15	1	IW
Sagittaria latifolia (arrowhead)	tubers	swamps, ponds sloughs	occ.-common throughout	15	0	IW
Sambucus canadensis (elderberry)	seeds***+	woods	common throughout	11	1	IW
Sassafras albidum (sassafras)	seeds+	dry or moist	common southern three-fourths	5	-	U
Scirpus spp. (bullrush)	seeds***+ tubers	shallow water	common throughout	17	?	U, IOR
Shepherdia canadensis (buffalo berry)	seeds***+	shores L. Michigan	rare	?	-	IOR, U
Smilax spp. (greenbrier)	seeds***+	woods, thickets	occ.-common throughout	15	-	U
Solanum americanum	seeds***+	disturbed	occasional	?	-	IW?
Staphylea trifolia (bladdernut)	seeds+	moist woods	common throughout	11	-	U
Strophostyles helvola (wild bean)	seeds***+	rocky woods	occasional	7	-	IW
Typha spp. (cattail)	roots pollen	wet grounds	common throughout			IW?

Table 7 (Concluded).

SPECIES	PART	HABITAT	ABUNDANCE	NH	YBC	STATUS
Vaccinium spp. (blueberry)	seeds***+	sandy woods, cliffs, bogs	occ.-rare	4	-	IOR,U
Verbena hastata (blue vervain)	seeds***+	moist ground	common throughout	8	-	U
Viburnum spp. (nannyberry, cranberry)	seeds***+	woods	occ. throughout	3	-	IW
Vitis spp. (grape)	seeds***+	woods, bluffs bluffs	occ.-common throughout	22	1	IW
Xanthium commune (cocklebur)	seeds+***+	waste ground	common throughout	5	-	U
Zea mays (corn, maize)	seeds***+ cobs***+					C
Zizania aquatica (wild rice)	seeds ***+	shallow water	local	-	-	IW

*NH=number of habitats of occurrence.
**YBC=average number of years between good crops.
***Remains have been found archaeologically.
+Potentially preservable parts are present.
#Status based on Historic usage and archaeological evidence: C=cultivated plant, IOR=important plant food in other geographical regions, IV=important wild plant food, Q=quasi-cultigen, W=widely used by Historic Indian groups, U=apparently unimportant in this region.

Illinois. These plants were used historically by various groups within their geographical range. Their absence, thus far, from the archaeological record of this area may be accidental, and some of the plants on this list, or closely related species, may yet emerge as likely food plants of Native Americans in Illinois.

Finally, a series of cultigens and possible cultigens that have been found in archaeological sites in Illinois are included. Most of these plants are introduced species rather than native, and their presence alone indicates that they must have been cultivated. We are most interested in when these plants were introduced or domesticated and how important they may have been to human subsistence in Illinois during various cultural periods.

ENVIRONMENTAL FACTORS AFFECTING HUMAN SUBSISTENCE

There are a number of environmental factors present in Illinois that may have affected human subsistence. Both the number and relative percentage of fruit-producing species are highest in the north while acorns and nuts are represented by the most species in the south. Only in northwestern Illinois and in the Forest-Prairie Floristic Province are there more nut than acorn species represented. This suggests the possibility that, simply because of higher diversity and presumably abundance, fruit may have been a potentially more important food resource in the north while nuts and acorns were more important in the south.

The generally high indices of similarity between the food plants of counties throughout Illinois suggest that extensive changes in subsistence would probably not be necessary for a hypothetical aboriginal group moving about the state. Even in the Grand Prairie Province, enough forest occurs along stream valleys and other rugged topographic settings that the index of similarity with the Northeastern Forest is 60. For a small group, it probably would not be difficult to find familiar foods. However, the abundance of specific plant foods might vary greatly, and larger groups might be obliged to shift to more abundant local resources.

Aboriginal groups traveling the length of Illinois via the Mississippi and Illinois rivers would spend a maximum amount of time in transition zones. Not only would water be the easiest mode of transportation but both the diversity and the homogeneity of plant resources along the route would be near maximum.

EFFECTS OF CLIMATIC CHANGE

Although far more difficult to substantiate, there appears to have been a relationship between climatic changes and changes in human subsistence activities at certain times in the past. This is not to imply that all cultural changes were, in fact, determined by climate because this is certainly not the case. However, climatic changes during the Holocene altered the distribution and composition of biotic communities and must, to some extent, have affected human subsistence and other aspects of culture.

The climatic fluctuations during the Holocene have included both a major event, the Hypsithermal warm/dry period lasting at least 3000 years, and a series of minor oscillations which are perhaps best understood for the last 3000 years (J. King 1981, Bernabo 1981). During the early Holocene, human populations of the Midwest were small as was the diversity of food plants utilized, and niche width was narrow (Christenson 1981). The relatively large-scale climatic changes of the mid-Holocene caused significant alterations in the distribution and composition of most plant communities. Many upland sites went from mesic forest to dry forest or prairie (J. King 1981); these changes must have affected human subsistence. For example, evidence from Rodgers Shelter in the western Missouri Ozarks shows a shift during this period to a diffuse, foraging economy paralleling a significant decline in deer hunting possibly brought about by prairie replacement of forest and a resulting decline in the deer population. At Graham Cave in central Missouri, there was a corresponding shift from emphasis on forest-adapted species to forest-edge species, reflecting the opening of the forest canopy (McMillan and Klippel 1981).

During the late Holocene (after 5000 YBP), climatic fluctuations were minor. Human culture during this period was increasingly complex. Resource diversity was high and niche width wide but diminishing with the increasing importance of plant cultivation. The effects of environmental fluctuations must have been lessened by availability of cultivated plant foods and by increasingly structured society.

The major problem with attempted correlations of climatic and cultural changes is that both are independently time transgressive. For example, prairie was established by 9900 YBP in northeastern Kansas but did not replace oak-hickory forest in the central Illinois uplands until about 8300 YBP. Stoltman (1978) lists eight separate definitions of an Archaic "period" beginning from 8000 B.C. to 1000 B.C. and ending from 4500 B.C. to ca. A.D. 350 that archaeologists have used to "designate an increment of time whose precise temporal limits vary with the specific geographic locality or region being considered." It is obvious that any person attempting to prove, or disprove, connections between climate and culture must be extremely careful to use the best temporal data available for the geographical area being studied and to stay within the limits of those data.

ARCHAEOLOGICAL EVIDENCE FOR PLANT UTILIZATION

The exact archaeobotanical flora preserved in a site depends on several factors including site size, duration and season of occupation, resources utilized, deposition and erosion rates, and preservation environment (pH, soil type, amount of moisture). Upland campsites tended to be small, short-term, special purpose encampments while larger, longer-term habitations were generally located near navigable streams. During the Middle Woodland and Mississippian periods (Table 8), populations were centered in the major river valleys while at other times they were more dispersed. Larger sites tend to have more well-defined middens and other features, including large storage structures in later sites, which may promote preservation of plant remains while also encouraging excavation and flotation.

TABLE 8. General Prehistoric and Historic Period Cultural Chronology for the Midwestern United States

PERIOD	YEARS B.C.-A.D.	YEARS B.P.
Prehistoric		
Paleo-Indian	10000-8000 B.C.	12000-10000 B.P.
Archaic		
Early Archaic	8000-6000	10000-8000
Middle Archaic	6000-2550	8000-4550
Late Archaic	2550- 600	4550-2600
Woodland		
Early Woodland	600- 200	2550-2150
Middle Woodland	200 B.C.-A.D. 300	2150-1650
Late Woodland	300-1180	1650- 770
Mississippian	900-1350	1050- 600
Oneota	1350-1573	
Historic	1673-1832	277- 118
Euro-American	1673-present	277-present

Source: Based on Wiant 1983: Table 8-1.

The analysis of plant remains has evolved in the last few decades as increasingly sophisticated techniques have been applied to increasingly complex research questions. In many early excavations, plant remains were limited to those which were noticed and removed by hand or those large enough to be caught by screens with mesh sizes of one-quarter inch or larger. The advent of water flotation (Struever 1968a) and subsequent improvements (Watson 1976 and many others) greatly increased the amount of subsistence data recovered.

The research interests of those analyzing archaeobotanical remains have also changed through time. Much of the early work on plant remains from Illinois sites was done by Dr. Hugh C. Cutler at the Missouri Botanical Garden. Cutler and his later associate Leonard Blake have, between them, over 130 publications, one of the most notable of which is their joint publication *Plants from Archaeological Sites East of the Rockies* (Cutler and Blake 1976). Dr. Cutler's primary interest was in the history of cultivated plants rather than in human subsistence or man-land relationships, and one of the first integrations of human subsistence data is presented by Munson *et al.* (1971) for the Scovill Site in Fulton County. Presently, a relatively large number of archaeobotanists, or paleoethnobotanists, are involved in studies of prehistoric human plant usage at various sites in Illinois.

The development of fine-screening and water flotation techniques greatly altered ideas about prehistoric human plant use. Large-mesh screen recovered only large, durable remains such as wood charcoal, nuts, and corn cobs and kernels while small or fragile remains were generally either destroyed or lost. Thus, it was originally felt that prior to the Mississippian Period, hunting and nut-gathering were the mainstay of the aboriginal diet (Caldwell's *Primary Forest Efficiency*) although there were pockets of "home-grown horticulture" or other localized adaptations (Caldwell 1958:18). Struever (1968b:305) applied Caldwell's concept that subsistence patterns were modified by "specific adjustments to regional resources" to partially explain the large quantities of small, wild seeds (sumpweed, knotweed, and chenopod) he was then finding in archaeological sites in the Lower Illinois Valley. Subsequent work there has shown that not only these but several other plants were being exploited as well (Asch and Asch 1978a, 1978b, 1981) while research in other regions has further demonstrated that, instead of being localized, use of certain of these plants was actually widespread in eastern North America (Yarnell 1978, Ford 1981).

There is a continuum in human plant usage stretching from mere exploitation no different from that of other animals on one extreme to domestication and concomitant genetic changes which make the plant an obligate cultigen on the other. Between these two extremes are varying situations in which plant populations could be accidentally enhanced by human activities such as disturbance, tended in some manner such as by weed removal, cultivated by tilling, or the seeds stored and dispersed (Ford 1979).

CULTIVATED PLANTS

A "cultigen" is one of a relatively small number of plant species or varieties not known in a wild form and therefore presumed to have originated through domestication (e.g., corn). Some cultivated plants evolve into cultigens as a result of human selection; others remain phenotypically, if not genotypically, unaltered.

Asch and Asch (1982) list three criteria for recognizing prehistoric cultivation including morphological changes indicating domestication, occurrence outside the natural range, and evidence for a level of utilization above that which could be sustained by exploiting natural stands. Unfortunately, most of the plants used by aboriginal peoples in eastern North America were either never domesticated or domesticated relatively recently. In addition, many economically important species are widely distributed. The initial usage of a specific plant would also start off on a small scale, probably indistinguishable in the archaeological record from naturally occurring specimens that were accidentally carbonized. Therefore, throughout most of the archaeological record the exact nature of exploitation of a given species is problematical. Recognizing these problems, Yarnell (1983) uses the term "quasicultigen" to imply some degree of cultigen status (short of domestication) and notes that "each crop plant is largely an independent entity in its space and time distributions...." The evidence appears adequate to say that, in addition to cultigens, various other plants were cultivated to the extent of being quasicultigens in some areas of Illinois such as the Lower Illinois River Valley and the American Bottoms. However, evidence for plant cultivation is lacking for many other portions of the state.

Three plants, chenopod, knotweed, and maygrass, have been repeatedly found together in Illinois archaeological sites. They sometimes occur in enormous quantities, suggesting similar levels of cultivation and use in this area. The present range of maygrass (*Phalaris caroliniana*) does not include Illinois, and there is no evidence that it grew here naturally in the past (Cowan 1978). Nutritional studies (Crites and Terry 1984) indicate that it was at least equal to corn as a food resource. Maygrass first appears during the Late Archaic in the Lower Illinois Valley (Asch and Asch 1981). It has been recovered as far north as the Late Woodland Rench site on the Illinois River near Peoria (F. King 1983) and is common in Middle and Late Woodland sites in the Lower Illinois Valley (Asch and Asch 1982) and American Bottoms (Johannessen 1981, 1983a, 1983b, 1984).

Asch and Asch (1977, 1982) identify the *Chenopodium* from the Lower Illinois Valley as *C. bushianum* and note that it has been recovered from late Pleistocene sediments in the area and archaeological contexts predating 5000 YBP Wilson (1981), on the other hand, argues that *C. bushianum* is much younger and probably derived from *C. berlandieri* var. *nuttalliae*, a chenopod of the Mexican highlands which he has identified in material from the Ozark bluff shelters. Whatever its origin, *C. bushianum*-type seeds increase in number substantially after about 4000 years ago in the Lower Illinois Valley. Asch and Asch (1982) suggest it was cultivated in this locality by 2000 YBP because of the restricted distributions of natural stands and some differences in seed morphology seen in specimens from two Late Woodland sites dating about 1300 YBP Although *carbonized Chenopodium* seeds are common in Illinois archaeological sites, preservation is often inadequate to permit identification to the species level, and the number of sites from which *C. bushianum*-type seeds have been reported is relatively small. Judgments regarding the exact species of *Chenopodium* involved and their status as possible cultigens will have to await further research. *Chenopodium bushianum*-type seeds have also been identified from Middle Woodland through Mississippian components in the American Bottoms (Johannessen 1984).

Knotweed, *Polygonum erectum*, was first identified to species from the early Late Woodland Scovill site by Munson *et al.* (1971) who felt that the species, which grows in disturbed soil, might have been an encouraged volunteer or even cultivated to some extent without intentional planting. Because of its abundance and regular occurrence with maygrass and goosefoot, it may have been cultivated during the Middle and Late Woodland in the Lower Illinois Valley (Asch and Asch 1982). Knotweed also occurs with goosefoot and maygrass at the Late Woodland Rench site near Peoria (F. King 1983) and in Middle Woodland through Mississippian components in the American Bottoms (Johannessen 1983a, 1983b, 1984; Whalley 1983), and in southern Illinois (Lopinot 1981a, 1981b, 1982). Although it has been recovered from central and eastern Tennessee (Yarnell 1983), its prehistoric use seems to have been limited in geographical distribution. *P. erectum* is the most common knotweed in the archaeological record although numerous other species have been found in Illinois sites as well. Considering the great physical similarity between the seeds of various species, it seems probable that they may also have been used under some circumstances.

Sumpweed, *Iva annua* var. *macrocarpa*, is an extremely interesting plant because it is one of the few native domesticates in eastern North America and apparently the only one to become extinct, possibly in early Historic times. Throughout the archaeological record, there is a gradual increase in size of approximately one mm/ millennium from an average length of approximately 3.0 mm for wild plants up to 13 mm for achenes recovered from Ozark bluff shelters and 9 mm from Mississippian sites in southeastern Missouri (Yarnell 1972, 1977, Asch and Asch 1978a). Sumpweed appears to be native from southern Illinois southward to Mississippi and westward along river valleys into the Great Plains (R. Jackson 1960, Black 1963). Archaeological evidence indicates that it was first domesticated in the area of Kentucky, southern Illinois, and eastern Missouri and reached its greatest distribution during the Mississippian when it was grown from western Arkansas and Iowa to western North Carolina (Yarnell 1978). The first appearance of sumpweed in the archaeological record is the occurrence of a achene in a Middle Archaic horizon at the Koster site dated approximately 7300 YBP (Asch and Asch 1978a). Intermediate-sized achenes from a 4000 YBP Titterington phase pit at the Napoleon Hollow site provide the earliest evidence of its domestication. Achenes from early Late Woodland sites in the Lower Illinois Valley average significantly larger than modern specimens (Asch and Asch 1978a, 1982). Elsewhere in Illinois, wild-sized seeds occur in Late Archaic features in the vicinity of the Cahokia site in St. Clair County (Lopinot 1983), at the early Late Woodland Griffie site in Mason County (F. King and Roper 1976), and the early Late Woodland Scovill site in Fulton County (Munson *et al.* 1971). Cultigen-sized specimens have been recovered from early Late Woodland and Mississippian features in the American Bottoms (F. King 1980, Johannessen 1983c, 1984, Johannessen and Whalley 1981).

Unlike sumpweed, sunflower (*Helianthus annuus*) is not native to eastern North America but to the western United States. Interestingly, however, most of the archaeological evidence for domestication has come from Kentucky, Tennessee, and adjacent areas. By early Historic times, sunflower was grown from eastern Texas to North Carolina, Quebec, and North Dakota although it seems to have been less important as a crop plant than sumpweed in Illinois and Missouri (Yarnell 1978:290-91, 1983). There are numerous wild species of sunflower in Illinois with extremely similar seeds, and wild-sized specimens from Illinois archaeological sites frequently cannot be identified to species (Asch and Asch 1982, Johannessen 1981, Whalley 1983). A single carbonized achene from the Late Archaic Pabst site measures 4.6x3.1 mm (6.0x4.3 mm with correction) and falls into the range of modern ruderal sunflowers (Letter from the author to R. B. Lewis, 2 March 1979). A seed from the Riverton site measures 6.8x4.0 mm. and dates approximately 3200 YBP (Yarnell 1977). *H. annuus* has been identified from emergent Mississippian and Mississippian components in the American Bottoms (Johannessen 1984).

Current evidence in the form of rind fragments from the Koster and Napoleon Hollow sites suggests that a species of *Cucurbita*, possibly *C.pepo*, was present in the Lower Illinois Valley by about 7000 years ago (Asch and Asch 1982, Conard *et al.* 1984). *Cucurbita* becomes common in archaeological sites in the Lower Illinois Valley during the Middle Woodland although its presence cannot be verified in the American Bottoms before Late Woodland (Asch and Asch 1982, Johannessen 1983c, 1984, Johannessen and Whalley 1981, F. King 1980). It has also been recovered from the early Late Woodland Scovill site in Fulton County (Munson *et al.* 1971). Although bottle gourd (*Lagenaria siceraria*) appears in eastern North America by at least 4500 years ago (Kay *et al.* 1980), the only known specimens recovered thus far in Illinois are from the early Late Woodland Newbridge and Carlin sites in the Lower Illinois Valley (Asch and Asch 1981) and from under Mound 51 at Cahokia (Cutler and Blake 1976).

Watermelon (*Citrullus vulgaris*) was introduced into the southeastern United States during the sixteenth century and rapidly spread throughout North America. In Illinois, it has been recovered from the Historic Zimmerman and Rhoads sites (Blake and Cutler 1975, Blake 1981).

Maize or corn (*Zea mays*) is an introduced tropical cultigen that assumed primary dominance as a food resource during the Mississippian Period. It was at one time felt to have also been instrumental in the rise and fall of the Hopewellian culture of the upper Great Lakes area (J. Griffin 1960), although increasingly sophisticated excavation and radiometric techniques have pushed the appearance continually later. Yarnell (1983) feels that corn probably did not arrive in eastern North America until about 1650 YBP The linear accelerator dating of suspected Archaic and Middle Woodland maize from the Koster, Napoleon Hollow, and Jasper Newman sites has shown it is questionable that maize was present in the Lower Illinois Valley prior to 1370 YBP (Asch and Asch 1982, Conard *et al.* 1984). In the American Bottoms, maize is virtually absent from early Late Woodland sites, appearing first in Late Bluff/emergent Mississippian components (Johannessen

1982, 1983c, 1984). Elsewhere in Illinois, maize is relatively widespread and abundant in late Late Woodland and Mississippian sites (e.g., Cutler and Blake 1976).

Beans (*Phaseolus vulgaris*) first appear in eastern North America, and in Illinois, about 1000 years ago. They have been recovered from a number of middle and late Mississippian and historic sites in Illinois (Cutler and Blake 1976).

Tobacco (*Nicotiana rustica*) is uncommon in the archaeological record because of the tiny seeds and the apparent care with which this ceremonial plant was treated. Nonetheless, tobacco has been recovered from Late Woodland components in the American Bottoms where it predates maize by 200 to 300 years (Johannessen 1984, Whalley 1983).

Little wild barley (*Hordeum pusillum*, identified in some publications as "cf. *Festuca*") has been identified from Late Woodland sites in the lower and central Illinois valley (Asch and Asch 1982, F. King 1983) and emergent Mississippian components in the American Bottoms (Johannessen 1981, 1984). Although the evidence is still scanty, little wild barley seems to have been nearly as common as maygrass, knotweed, and goosefoot in some Illinois sites following its initial appearance during the early Late Woodland. The confusion that has accompanied the identification of carbonized and sometimes misshapen specimens of maygrass and little wild barley is readily understood. Even fresh grass caryopses are often difficult to identify. In addition to these grasses, several others occur in archaeological sites which remain unidentified. It is possible that some of these also represent food resources and that some may even have been locally encouraged or cultivated.

Giant ragweed (*Ambrosia trifida*) is an extremely common disturbance plant that probably falls somewhere between being an unused weed around aboriginal camps and a cultivated food plant. Gilmore (1931) suggested that it was a domesticate of the Ozark Bluff-Dwellers while Payne and Jones (1962) argue that it was not. Based on its abundance in Archaic components at the Koster and Napoleon Hollow sites, Asch and Asch (1982) speculate that in west-central Illinois it was "once a cultivated plant but that cultivation was abandoned prior to Woodland times."

Jerusalem artichoke (*Helianthus tuberosus*) is included by Yarnell (1983) among the quasicultigens. Although the plant is common on moist ground throughout Illinois, identifiable remains of the edible tubers have not been recovered from archaeological sites in the state.

WILD PLANT FOODS

In addition to the cultigens and quasicultigens, there is evidence for the importance of numerous wild plant foods. Yarnell (1983) lists amaranth, purslane, pokeweed, carpetweed, and sorrel as important sources of potherbs. A "minor complex of indigenous seed types" that includes wild bean (*Strophostyles* sp.), black nightshade (*Solanum* cf. *americanum*), and two grasses, one of which closely resembles little wild barley, occurs regularly in early/late Bluff transition and later assemblages in the American Bottoms (Johannessen 1981, Johannessen and Whalley 1981). Wild bean has also been recovered from numerous sites outside of the American Bottoms, including the middle Archaic horizons at the Modoc Rock Shelter (F. King 1981) and the Historic Zimmerman site (Asch and Asch 1975). Groundnut (*Apios americana*) was an important food plant of Historic Indians and, considering that tubers are generally poorly preserved, their presence in at least four sites may be significant. American lotus (*Nelumbo lutea*) another important food plant of Historic Native Americans, has been reported from Late Archaic through Historic contexts.

Nut remains are, of course, practically ubiquitous in archaeological sites, with the relative importance of hickory, pecan, black walnut, hazelnut and acorns varying through time and geographical location. The wild fruits most often recovered include blackberry/raspberry/dewberry (*Rubus* spp.), wild plum and wild cherry (*Prunus* spp.), grape (*Vitis* spp.), persimmon (*Diospyros virginiana*), and pawpaw (*Asimina triloba*). Occurring less frequently are red mulberry (*Morus rubra*), black haw (*Viburnum* spp.), shadbush (*Amelanchier* spp.), black huckleberry (*Gaylussacia baccata*), wild crabapple (*Malus* spp.), hawthorn (*Crataegus* spp.), and elderberry (*Sambucus canadensis*). There is good historical evidence for the use of all of these taxa for food.

As more data emerge, it becomes increasingly apparent that there are distinct, if minor, spatial and temporal variations in aboriginal plant use in Illinois. *Cucurbita* appears and becomes important much earlier in the Lower Illinois Valley than elsewhere.

Although widespread, small, starchy, or oily seeds vary in apparent abundance between sites, perhaps reflecting local environmental suitability for their growth, seasonality, human preference, the availability of other resources, and/or site function. Likewise, there are differences in the proportions of various types of nuts which may reflect environmental change through time, human preference, or processing procedures.

Some of the best archaeobotanical information comes from deeply stratified sites with long-term records of human occupation. Unfortunately, such sites are almost entirely restricted to the floodplains of the major river valleys. There probably were significant differences between the life-ways of inhabitants of the major river valleys and inhabitants of camps on small tributary streams or uplands. Although a good understanding of human subsistence is being developed for portions of the Illinois and Mississippi river valleys, much of the remainder of the state is either distressingly barren of archaeological remains or has not been studied. Although the answers may prove difficult to assemble, there are many questions left concerning prehistoric plant use in Illinois.

Current evidence suggests that Illinois Indians existed primarily on wild plant and animal foods prior to the Late Archaic Period. Various types of nutshell and wood charcoal are the most common types of plant remains found in archaeological sites dating prior to this period. Exotic domesticated plants (squash and bottle gourd) were apparently introduced rather early, possibly by 7000 years ago (based on material from the Koster and Napoleon Hollow sites reported by Asch and Asch 1982 and Conard *et al.* 1984).

The first occurrence of native plant species that were ultimately either cultivated or domesticated is that of wild-sized achenes of sumpweed at the Koster site in a horizon dating 7500 YBP Other starchy seeds also occur in small numbers in Late Archaic samples from this site. The oldest sunflower thus far recovered with a seed size suggesting the possibility of early cultivation comes from the Riverton site in Crawford County and dates to about 3200 YBP (Yarnell 1977).

The patterns established during the Late Archaic apparently continued during the Early Woodland with low level cultivation or usage of several introduced or wild species. However, by Middle Woodland times, a number of plant species may have been commonly cultivated or at least "encouraged" including chenopod, knotweed, maygrass, sumpweed, and sunflower. Little wild barley first appears during the early Late Woodland.

During the Late Woodland, small seeds gain in importance while nut utilization apparently declined and became increasingly specific for hickory nut. Current evidence suggests that maize first arrived in Illinois sometime during the late Late Woodland/ early Mississippian. The Mississippian peoples had large population centers which were heavily dependent on plant cultivation, particularly corn, although beans, squash, sunflower, sumpweed, and other plants were also important.

HISTORIC PLANT UTILIZATION

DEFINITION OF ETHNOBOTANY

Ethnobotany is the study of the uses of plants for various purposes by particular ethnic groups. Although the term "ethnobotany" was first formally used in 1895 (Harshberger 1896), European scholars had long been interested in the history of domesticated plants and the vegetal remains found in archaeological sites in the Old World (Ford 1979). The early explorers of the New World noted with eager enthusiasm the new (to them) plants the Indians were using. Tobacco, for example, was an immediate success in Europe. During the early twentieth century, detailed ethnobotanies were published for several North American Indian tribes that were rapidly disappearing or being acculturated. Although ethnobotanies continue to be written for various contemporary groups (e.g., Black 1980, Bolyard 1980, Grime 1979), the use of wild plants is being increasingly altered everywhere by the introduction and acceptance of foreign plant species.

The first Euro-American settlers in Illinois brought with them the seeds of many cultivated plants to be used for food, medicine, and ornament. Many of these plants were quickly adopted by the Native Americans. Watermelons, peaches, cucumbers, and new types of squash spread rapidly among Indian groups after their initial introduction. However, the addition of each new cultivated food plant to the diet meant that traditional wild plants were probably used less frequently. Rapid acceptance of new plants coupled with an oral rather than a written history meant that the use of many other plants was gradually forgotten. For example, Waugh (1916) states that some wild and formerly cultivated plants were "still eaten by some of the older people" while others "have been practically forgotten by the present-day Iroquois."

COMPILATION OF ETHNOBOTANICAL INFORMATION

Ethnobotanical data are compiled from many sources, the most important of which is the work of researchers who have spent time with specific groups of people learning their uses for plants. Such studies have been compiled for many groups and continue to be published today.

Other sources include descriptions by early missionaries, travelers, and settlers. The accounts of the early Jesuit missionaries contain valuable descriptions of many plants. The usefulness of such information unfortunately varies with the botanical knowledge of the observer. At times, the description is completely inadequate to identify the plant being discussed. In some cases, the root or flower of one plant is associated with the leaves of another, making an impossible combination. In addition, the early travelers often tended to romanticize or exaggerate the merits of the new territory, including such things as the size of fruits or tubers. Even the early ethnobotanies written by experienced students of plant usage contain inconsistencies and errors reflecting the limits of the ethnobotanist's taxonomic knowledge and the difficulty in identifying unfamiliar fragmentary plant specimens. Sometimes too, the ethnobotanist was accidentally or deliberately misled by an informant about a plant's use, habitat, or identification.

A final source of ethnobotanical information is the archaeological record. Carbonized plant remains resulting from the accidental burning of foodstuffs or the use of plant materials for fuel are present in many sites. However, the reconstruction of plant use on the basis of such data is difficult. Not all plant parts preserve equally well, and prehistoric preparation techniques bias the types of remains that are likely to be carbonized and thereby preserved. Furthermore, the possibility exists that plants with no economic significance might be accidentally incorporated into site deposits because they were growing on the site, were attached to firewood or clothing, or for some other reason. Only when a plant occurs repeatedly at many archaeological sites is it safe to assume that it was being used for food or some other economic purpose.

The Historic Indian groups of Illinois disappeared before they were thoroughly studied by ethnographers. Therefore, this compendium of the useful plants of Illinois is based on the ethnobotanies of other Algonquin-speaking groups to the North and East, as well as groups to the West and South. For that reason, not all of the species included here were necessarily used by the Indians of Illinois, nor were they all used by any other single aboriginal group.

The distribution of plant and animal communities varies over even relatively small distances. Certain plants used for food, medicine, or other purposes may be common in one area and rare or absent in another. Some plants are recorded as having been used by virtually all of the Indian groups within their geographical range; others are mentioned as having been used by only a single group. Occasionally, dubious plants or usages occur in the literature which may be due to mistaken identifications on the part of the researchers recording plant usage, lack of familiarity with the local botany, or errors in translation of Indian dialects. Where such mistakes appear likely, the more probable plant species has been indicated.

The specific group of plants used by a group was the result of many factors. These include the length of time a group had occupied an area thus gaining familiarity with the local flora; the type and abundance of potentially useful plants that were available; seasonal activity scheduling which might promote the collection of one plant over another; cultural preferences and prejudices, and the availability of cultivated plants. It is no wonder, therefore, that the plants used might differ substantially, even between groups that were closely related or in close geographical proximity.

The distribution of plants in Illinois today is not the same as that of 10,000, 5,000, or even 1,000 years ago. Climatic change, plant migration, fire, and human activity have contributed to and altered the composition of the biotic communities that have existed at various times in the past. The role of prehistoric peoples and the plants they used to support their lifeways cannot be separated from paleoecological events of the same region.

ILLINOIS ETHNOBOTANY

The following list includes all of the vascular plant species that may be encountered growing "wild" in Illinois and that are known to have any past or present human use. In order to be of value to the maximum number of users, this list includes all plants that occur in the state whether they are native or introduced, abundant or rare. Because of the climatic changes that have occurred since people first arrived in North America and because of more recent large-scale changes of the landscape by modern human populations, plant communities are much different today than those that existed in the past. Many species that are rare or extinct in Illinois today, particularly plants adapted to the cooler and moister conditions of the north, were once much more common. Plants that are rare are included in this list with the assumption that they may once have been more common and widespread and therefore of greater potential economic value than those that have disappeared entirely.

The usages on the following list were compiled from the ethnobotanies of numerous Indian groups over a relatively large geographical area. It must be emphasized that not all of the native plants on this list were used by the Indians of Illinois. In any one region, many plants share the same potential function for food, fiber, or medicine; and the plants used by the Indians were undoubtedly limited to a relatively small number of those occurring in their area. The reasons that a certain plant might be selected over another are numerous: food plants for their abundance, food value, taste and ease of collection and preparation; medicinal plants for their accessibility and perceived effectiveness. The specific plants used for dyes, technological purposes, and smoking also probably depended on a balance between desirability and availability.

Plant usage has been compiled primarily from the sources listed in the references for each plant. The most important ethnobotanies used include Densmore (1928), Gilmore (1919), H. Smith (1923, 1928, 1932, 1933), Swanton (1946), and Waugh (1916). An extremely valuable synthesis of data on plant uses by Indians of the Great Lakes Region is that by Yarnell (1964).

Plant Foods

Probably the most important aboriginal use of plants was for food, and the distribution of plant food resources played an important role in determining human settlement and subsistence patterns. Likewise, changes in the availability of various plant resources due to environmental shifts must have influenced human activities. In addition to the primary ethnobotanical sources listed above, Yanovksy (1936) presents a very useful compilation of Indian food plants. Zawacki and Hausfater (1969) describe the early vegetation and potential plant food resources of the Lower Illinois Valley.

Medicinal Plants

A great many species of plants were used for medicinal purposes by one Native American group or another. However, as noted earlier, no single historic group used all the plants listed here. Of those that were used, some were considered particularly effective while others were used only when nothing better was available. The use of a certain plant for a given illness must have been arrived at, in part, through much empirical experimentation. If a certain plant were used and the patient survived, the plant would probably be used again for the same or a similar illness. If the patient grew worse or died, the search would continue. Many plants appear to have been used to treat conditions for which they must have had little efficacy. Factors such as a person's "will to live" and a strong belief that he or she would be cured must also have been important.

Plant usage by Native Americans is indicated by references to the tribe or by "Indian" if the names of tribes using the plant are unavailable; for example, references sometimes note only that a plant was used "by Indians in eastern North America." In addition, Lewis and Elvin-Lewis (1977), Tehon (1951) and Vogel (1970) were very helpful in finding more recent Euro-American uses. Lewis and Elvin-Lewis (1977) describe past and present medical botany in an extremely interesting and well-written textbook format. Tehon's publication would be of value to anyone interested in uses of medicinal plants in Illinois prior to the proliferation in modern "miracle drugs"; it describes the methods of collection and preservation of plants sold to drug manufacturers at that time. Only a few of the drug plants included by Tehon were listed as "official" at the time in the *Eleventh Decennial Revision of the Pharmacopoeia of the United States* (and even fewer are official today). However, many had been listed in earlier editions of the *Pharmacopoeia* or in the *National Formulary* and most were still used to varying extents as the source of drugs or in candies, flavorings, and perfumes. Herbal medicines today are confined almost entirely to "health food" stores where they are purchased for self-medication, and many modern herbals are available promoting just such use. Because the use of herbal medicines has not disappeared, the term "recent" in the following listing is used to denote all Euro-American uses including a few continuing into the present. Vogel (1970) includes not only plants used by Native Americans but those relied upon by early Euro-American settlers and would be of interest to persons studying the medicines of that era. Harn and Koelling (1974) list the indigenous drug plants found in Fulton County.

There are many problems involved in the identification and documentation of plant use. The problems arise from lack of familiarity with the native American flora on the part of the early European-trained botanists and ethnobotanists, as well as language barriers, cultural barriers, and the difficulty of attempting to identify unfamiliar fragmentary and dried specimens. An excellent discussion of documenting and evaluating modern herbal remedies is given by Croom (1983).

Virtually everyone has some contact with such herbs every day, if nothing more than a little black pepper or nutmeg in one's dinner. Many herbs are used alone or in mixtures for herbal teas and contain active ingredients which act as stimulants, diuretics, tonics or in other ways that add to a feeling of well-being. For example, the most common stimulant is caffeine which occurs in coffee, tea, cocoa, and many soft drinks. Many commonly cultivated garden herbs also act as mild stimulants (chamomile, yarrow, fennel, elecampane, horehound, mint, germander, garlic) as do many less familiar wild plants.

Perhaps as a result of some disillusionment with modern medicine, there is currently a renewed interest in the use of herbal remedies. And, as noted by Tyler (1982:1), "more misinformation regarding the efficacy of herbs is currently being placed before the consumers than at any previous time, including the turn-of-the-century heyday of patent medicines."

No person should ever attempt to treat a serious ailment with home remedies or herbal medicines instead of seeking proper professional help. There are good reasons that few of these plant-derived drugs remain official today. Many of them are really a combination of several more or less similar compounds produced by the plant, each having its own effect on the human body, frequently with serious side effects. Many of them are poisonous even in small amounts, making proper dosages difficult to calculate. Some can interact dangerously with other medicines. Despite a widely presumed superiority of anything "natural," many popular herbal medicines are either totally ineffective or much less effective than modern drugs specifically formulated for a given ailment.

Dye Plants

The following list includes dye plants used by historic Indians and also those discussed in several modern references.

Many of the natural plant dyes fade rapidly; also they often are rather dull "earth" tones of yellow, green, and brown. Red and blue were particularly difficult to achieve before the advent of modern aniline dyes. Indian dyers sometimes resorted to material such as poke berries which provide a bright, but short-lived red.

For those interested in natural dyes, there are many good references available. The handbooks by the Brooklyn Botanical Garden (1964, 1973) are especially useful. Other sources used in this compilation include Bliss (1978) and Lust (1974).

Craft and Ceremonial Uses

Indian uses of plants for craft and ceremonial purposes are probably not directly applicable to modern pursuits. Most historic tribes made heavy use of plants for hunting and fishing charms, love charms, charms to ward off evil spirits, to produce dreams, or to make dreams come true. Some of the plants used for "medicine" may well have been used for charms or "bad medicine."

Technological Uses

The plants used for technological purposes were often those best suited to a certain type of construction and often the same species as those we use today for similar items. Strong fibers from plants such as basswood or nettle were used for weaving fabrics, bags, and nets. Long, straight stems were used for arrow shafts; fragrant plants were used for bedding. Woven fabrics and carved utensils of fine craftsmanship were made early in prehistoric times but unfortunately are rarely preserved.

Smoking Materials

The leaves and bark of many plants were used for smoking material, either for pleasure, to attract game, or for ceremonial purposes. Even after the widespread acceptance of tobacco, wild plants were often included in smoking mixtures to extend a limited resource. Tobacco was frequently so valued that those tribes that had it would often trade it or use it as a gift; however, the seeds were carefully guarded to maintain the economic value of the plant. A valuable reference on plants used for smoking is Epstein (1981).

Terminology on This List

Mohlenbrock's *Guide to the Vascular Flora of Illinois* (1975) was selected as the basic botanical reference because most people identifying plants in the field would probably use this manual rather than Fernald (1950) which is larger and more general in scope but more difficult to use. In addition, the nomenclature of Mohlenbrock's guide is more current than G. Jones (1950) and is compatible with the distribution maps in Mohlenbrock and Ladd (1978), the description of the ecology of southern Illinois by Voigt and Mohlenbrock (1964), and Mohlenbrock's *Illustrated Flora of Illinois* series (1967–).

The following information is given for each plant on this list:

SCIENTIFIC NAME is that listed in Mohlenbrock (1975). A species name consists of two words. The first part of the scientific name is that of the genus followed by the specific epithet. Many species occur in the same genus but a species name is generally distinct to a group of virtually identical plants. Scientific names are in Latin, and the specific epithet is often an adjective describing some attribute of the species that distinguishes it from the rest of the genus. For example, the genus *Quercus*, to which the oaks belong, is very large but there is only one *Quercus alba* or white oak.

A species is sometimes further subdivided if it includes populations that differ significantly. Potential subdivisions of species include subspecies, variety, and form. These have generally been ignored in this compilation because it usually requires an experienced taxonomist with a large collection of reference specimens to distinguish the differences. Moreover, such distinction was probably rare among Native American herbalists and of even less interest to persons out collecting nuts or berries as a source of food. Exceptions might exist; for example, there might conceivably have been selection for varieties that had larger fruit or nuts such as *Carya glabra* var. *megacarpa* Sarg., the large-fruited sweet pignut.

AUTHORITY is the person who first published on a given taxon (e.g.,the "L." in *Quercus alba* L. is a standard abbreviation for Carolus Linnaeus). The authority is used because many plants have had several different scientific names through time as various taxonomists, working independently, name, rename, split, or combine the taxa. Generally, it is the first name given to a taxon that has become official, unless that name had been previously assigned to another plant. By listing the authority, it is known precisely which taxon is being discussed. When a taxon in renamed, the original authority is retained in parentheses (e.g., *Pilea pumila* (L.)Gray).

COMMON NAME is that listed by Mohlenbrock (1975). For many plants, additional common names are used in other geographical areas and can be found in Fernald (1950) or in various regional publications.

SYNONYMS are obsolete scientific names. Most plants have one or more synonyms. A complete listing for Illinois plants is given by Jones and Fuller (1955). To avoid confusion due to nomenclatural changes that have occurred during the last 50 to 100 years, all synonyms occurring in any of the sources used in compiling this list are included here.

GROWTH FORM includes the size and shape of a plant. Most plants can be classified as trees, shrubs, herbs, vines, or grasses and the inclusion of this information may help in the understanding of plant usages. For example, to know that grapes grow as vines with long, thin flexible stems makes it easier to understand how they might have been used as cordage. HABITAT includes the conditions under which a plant will grow, including moisture, soils, amount of light needed, and many other factors. Habitat can only be approximated here because some plants have a wide range of tolerances while others are very narrow. General habitat preferences, as listed by Mohlenbrock (1975), are shown for each plant.

Although many of the plants on the list are NATIVE, many others are INTRODUCED. They were often brought to this country because of their value for food, fodder, dye, ornament, seasoning, or medicine while some came as contaminants in the seed of other plants. Interestingly, the ethnobotanies used in compiling this list show that numerous introduced plants were rapidly adopted, even for medicinal uses, by Native Americans during the last century. FREQUENCY is the abundance of a given species. This varies for different habitats and is also difficult to quantify. The frequency of each species is estimated as abundant, common, occasional, local, and rare. The DISTRIBUTION of plants vary because of their tolerances for differing environmental conditions. Some tolerant plants occur commonly throughout the state, others grow only under very specific conditions and thus have extremely limited distributions. Rare plants with limited distributions were probably not important food plants; however, their rarity might have made them more valuable for medicines, charms, or ceremonial purposes.

NOTE:

The following section reports Native American and Euro-American uses of various plants. Some plants recorded as having been used for food or medicine are considered to be poisonous by recent authors and have been so noted.

Neither this author nor the Illinois State Museum as publisher makes any recommendations as to the use of any plant for food or medicine. Anyone interested in such use should be certain of the identification of the plants. Further information can be found in references such as:

Fernald, M. R., A. C. Kinsey and R. C. Rollins.
1958 *Edible Wild Plants of Eastern North America.* Harper & Row, New York.

Kingsbury, J. M.
1964 *Poisonous Plants of the United States and Canada.* Prentice-Hall, Englewood Cliffs, N.J.

Lewis, W. H. and M. P. Elvin-Lewis
1977 *Medical Botany: Plants Affecting Man's Health.* John Wiley & Sons, New York.

Tyler, V. E.
1982 *The Honest Herbal: A Sensible Guide to Herbs and Related Remedies.* George F. Stickley Co., Philadelphia.

LISTING OF VASCULAR PLANTS BY FAMILY

The following list is arranged alphabetically by family, genus, and species within Divisions.

DIVISION I: PTERIDOPHYTA

Rushlike, mosslike, or fernlike vascular plants without flowers; reproduction by spores.

EQUISETACEAE
Horsetail Family

Equisetum arvense L.
Common horsetail
Native herb; common throughout Illinois in fields and disturbed areas.

MEDICINE: Used by Ojibwa and Potawatomi for dropsy, kidney, and bladder ailments. Recent--used as a diuretic; found toxic to cattle and horses.
DYE: Recent--herb used for yellow, gray-green, and green.
CEREMONY: Used as a "charm" by the Ojibwa.
REFERENCES: Brooklyn Botanic Garden 1964:73; Gilmore 1933:122; Lewis and Elvin-Lewis 1977:26, 312; H. Smith 1928:272, 1932:368, 1933:56.

Equisetum hyemale L.
Scouring rush
Native herb; common throughout Illinois on shores, banks and roadsides.

MEDICINE: Used by the Meskwaki for gonorrhea and by the Menominee and Ojibwa for kidney trouble.
TECHNOLOGY: Stems used by the Ojibwa to scour kettles.
REFERENCES: H. Smith 1923:34, 1928:220, 1932:368,418.

Equisetum laevigatum A.Br.
Smooth scouring rush
Native herb; common in open moist sandy areas in the northern half of Illinois, occasional elsewhere.

SYNONYMS: *E. hyemale* var. *intermedium* A.A. Eat.
FOOD: Ground plants eaten in New Mexico.
REFERENCES: Yanovsky 1936:4.

LYCOPODIACEAE
Clubmoss Family

Lycopodium inundatum L.
Bog clubmoss
Native herb; occurs in bogs in the northeastern part of Illinois; rare.

MEDICINE: Recent--contains a fixed oil; spores used as dusting powder.
REFERENCES: Tehon 1951:75.

OPHIOGLOSSACEAE
Adder's Tongue Family

Botrychium virginianum (L.)Sw.
Rattlesnake fern
Native herb; common in dry or moist woodlands throughout Illinois.

MEDICINE: Used by the Potawatomi for "medicine" and by the Ojibwa for lung trouble, consumption, and insect bites.
REFERENCES: Densmore 1928:288; H. Smith 1932:377, 1933:67.

OSMUNDACEAE
Royal Fern Family

Osmunda cinnamomea L.
Cinnamon fern
Native herb; occurs locally in swamps and swampy woods throughout Illinois.

FOOD: Young fronds eaten by the Menominee.
MEDICINE: Used by the Menominee to promote lactation.
REFERENCES: H. Smith 1923:70,124.

Osmunda regalis L.
Royal fern
Native herb; occurs locally in swamps, woods, or on moist ledges throughout Illinois.

MEDICINE: Used by the Menominee for "medicine."
REFERENCES: H. Smith 1923:44.

POLYPODIACEAE
Fern Family

Adiantum pedatum L.
Maidenhair fern
Native herb; common in moist, shaded woods throughout Illinois.

MEDICINE: Root used by the Potawatomi to make a medicinal beverage to promote lactation, by the Meskwaki for stomach cramps, and by the Menominee for leukorrhea. Recent--used as a demulcent and pectoral.
CEREMONY: Used by the Potawatomi as a good luck charm.
REFERENCES: Lewis and Elvin-Lewis 1977:236; H. Smith 1928:197, 1933:122; Vogel 1970:304,317.

Athyrium filix-femina (L.)Roth
Lady fern
Native herb; common throughout Illinois in moist, open woods and on the borders of swamps.

SYNONYMS: *A. angustum* (Willd.)Presl.
MEDICINE: Root used by the Ojibwa and Potawatomi to make a medicinal beverage and by the Meskwaki for breast pains.
REFERENCES: H. Smith 1928:237.

Dryopteris cristata (L.)Gray
Crested fern
Native herb; occurs occasionally in the northern one-half of Illinois in low, moist woodlands.

MEDICINE: Used by the Ojibwa to make a medicinal beverage.
REFERENCES: H. Smith 1933:381.

Dryopteris marginalis (L.)Gray
Marginal wood fern
Native herb; occasional to common throughout Illinois in rocky woods.

MEDICINE: Recent--used as an anthelminthic and taeniafuge.
REFERENCES: Vogel 1970:304.

Matteuccia struthiopteris (L.)Todaro
Ostrich fern
Native herb; restricted to the northern third of Illinois; occurs in rich, moist woods.

SYNONYMS: *Onoclea struthiopteris* (L.)Hoff.
MEDICINE: Used by the Menominee as a poultice.
REFERENCES: H. Smith 1923:47.

Onoclea sensibilis L.
Sensitive fern
Native herb; common in moist woods or on low, open ground throughout Illinois.

FOOD: Rootstock used for food by the Iroquois.
MEDICINE: Rootstock used for "medicine" by the Ojibwa to treat caked breasts. Recent--has been found to cause poisoning in livestock.
REFERENCES: Lewis and Elvin-Lewis 1977:26; H. Smith 1932:382; Waugh 1916:118; Yanovsky 1936:4.

Pteridium aquilinum (L.)Kuhn
Bracken fern
Native herb, common in open woods and fields throughout Illinois.

SYNONYMS: *P. latiusculum* (Desv.)Hieron
FOOD: Young sprouts eaten by the Ojibwa. Note--this fern is poisonous and should not be eaten at any stage.
MEDICINE: Root used by the Ojibwa and Menominee to make a medicinal beverage. Recent--this fern is carcinogenic and mutagenic and has caused poisoning in livestock.
DYE: Recent--roots used to make yellow, young shoots used for yellow-green and gray.
REFERENCES: Brooklyn Botanic Garden 1964:26, 32, 94; Lewis and Elvin-Lewis 1977:26; Lust 1974:553; H. Smith 1932:408.

DIVISION II: SPERMATOPHYTA
SUBDIVISION I: GYMNOSPERMAE

A group of generally evergreen plants characterized by ovules borne on open scales, usually in cones, a lack of true vessels in the wood, and needlelike or scalelike leaves.

CUPRESSACEAE
Cypress Family

Juniperus communis L.
Common juniper
Native shrub; occurs on sandy soil in the northeastern portion of Illinois; rare.

SYNONYMS: *J. canadensis* Burgsdorf
MEDICINE: Berries used by the Potawatomi to treat urinary tract problems, used by the Ojibwa for asthma. Recent--dried berries and oil considered useful as a diuretic, emmenagogue, expectorant, and genitourinary antiseptic.
TECHNOLOGY: Bark used by the Ojibwa for weaving mats and building houses.
DYE: Recent--fruit used to produce brown.
REFERENCES: Brooklyn Botanic Garden 1964: 68, 70; Densmore 1928:290; Lewis and Elvin-Lewis 1977:360; H. Smith 1933:69; Tehon 1951:70; Vogel 1970:329-30.

Juniperus virginiana L.
Red cedar
Native tree; common throughout Illinois in woods and fields and on cliffs.

MEDICINE: Beverage made from leaves used by the Meskwaki as a convalescent and by Plains Indians for coughs; twigs burned, the smoke inhaled by Indians of the Missouri River region for coughs; plant used by the Ojibwa for "medicine." Recent--plant considered toxic to livestock; of possible use in cancer chemotherapy.
TECHNOLOGY: Bark used by the Ojibwa and Potawatomi for weaving mats and bags and by the Missouri River Indians in making shelters.
REFERENCES: Densmore 1928:290; Gilmore 1919:11; Lewis and Elvin-Lewis 1977:27, 133, 285, 306; H. Smith 1928:234; Tehon 1951:70; Vogel 1970:289-90.

Thuja occidentalis L.
Arbor vitae
Native tree; occurs on cliffs, bluffs, and in bogs in the northeastern portion of Illinois; rare.

MEDICINE: Medicinal beverage made from inner bark used by the Menominee to treat suppressed menses; leaf infusion used by the Ojibwa to treat coughs and headache.
TECHNOLOGY: Wood used by the Ojibwa for canoe ribs, toboggans, and spear handles; roots used to sew canoes; bark used for rope, twine, nets, and bags.
DYE: Used by the Ojibwa.
CEREMONY: Dried leaves used by the Ojibwa to exorcise evil spirits.
REFERENCES: Densmore 1928:293,386; Gilmore 1933:123; H. Smith 1923:46,76, 1932:380,422, 1933:70,122; Vogel 1970:273-5; Stowe 1940:12.

PINACEAE
Pine Family

Larix laricina (Duroi)K.Koch
American larch
Native tree; occurs in bogs in the northeastern portion of Illinois; rare.

FOOD: Tea made from the roots by the Ojibwa.
MEDICINE: Dried leaves used by the Ojibwa as an inhalant and fumigant and root tea used for burns and anemia; bark used by Menominee and Potawatomi as a poultice and bark tea used for inflammation. Recent--used as a laxative, tonic, and diuretic.
TECHNOLOGY: Roots used by the Ojibwa to sew canoes and for woven bags.

REFERENCES: Densmore 1928:290; Gilmore 1933:123; H. Smith 1923:45, 1932:378, 1933:69; Yanovsky 1936:4.

Pinus banksiana Lamb.
Jack pine
Native tree; originally occurred on sandy soil in northeastern Illinois; at the present time natural stands are extinct or nearly extinct in this state.

MEDICINE: Pitch used as a reviver by the Ojibwa and Potawatomi and by the Ojibwa as an ointment; leaves used as a fumigant; all parts used by the Menominee for "medicine." The needles are toxic to livestock in large quantities and may cause contact dermatitis in some persons.
TECHNOLOGY: Roots used by the Ojibwa, Potawatomi, and Menominee for canoe and other coarse sewing and by the Potawatomi for torches.
REFERENCES: Lewis and Elvin-Lewis 1977:27,83; H. Smith 1923:45,75, 1932:379,421, 1933:70,113,122.

Pinus echinata Mill.
Shortleaf pine
Native tree; occurs naturally on rocky soil in Randolph and Union counties; rare.

MEDICINE: Resin used by Indians in the Southeast for tuberculosis and stomach troubles. Recent--pine tar used as an antibacterial, irritant, parasiticide, and expectorant.
REFERENCES: Vogel 1970:236,347-48.

Pinus resinosa Ait.
Red pine
Native tree; occurs in dry woods in La Salle county; rare.

MEDICINE: All parts used for medicine by the Ojibwa; powdered leaves used as an inhalant and reviver by the Potawatomi.
TECHNOLOGY: Roots used for sewing canoes by the Ojibwa; pitch used for sealing canoes, roofs, etc.
REFERENCES: Densmore 1928:291; H. Smith 1932:379,421, 1933:70.

Pinus strobus L.
White pine
Native tree; occurs in woods in the northern one-fourth of Illinois, local.

FOOD: Young staminate catkins cooked by the Iroquois and Ojibwa.
MEDICINE: Inner bark used for poultices and chest pains by the Menominee; pitch used as a salve by the Potawatomi; used as a reviver or inhalant by the Ojibwa. Recent--inner bark used as a mild expectorant.
TECHNOLOGY: Pitch used for sealing canoes by the Potawatomi and Ojibwa.
REFERENCES: Gilmore 1919:123; H. Smith 1923:46, 1932:379,408,421, 1933:70,122; Tehon 1951:87; Vogel 1970:346-47; Waugh 1916:117; Yanovsky 1936:6.

TAXACEAE
Yew Family

Taxus canadensis Marsh.
Canada yew
Native shrub; occurs on wooded hillsides in the northern one-fourth of Illinois; rare.

MEDICINE: Bark, leaves, and seeds are poisonous. Branches were used by the Meskwaki in steam treatment for rheumatism, paralysis, and numbness; root was used by the Potawatomi as a diuretic and in treatment for gonorrhea and by the Ojibwa for "medicine."
REFERENCES: Densmore 1928:293; Gilmore 1933:122; Lewis and Elvin-Lewis 1977:28; H. Smith 1923:54, 1933:84.

SUBDIVISION II: ANGIOSPERMAE
CLASS I: MONOCOTYLEDONEAE

Plants with one cotyledon, usually with parallel-veined, oblong or linear-shaped leaves, and with flower parts in threes or multiples of three.

ALISMATACEAE
Water-plantain Family

Sagittaria cuneata Sheldon
Arrowleaf
Native herb; rare in the northern half of Illinois where it occurs in mud or water in sloughs and along waterways.

FOOD: Tubers eaten by the Ojibwa.
MEDICINE: Used by the Ojibwa for indigestion.
REFERENCES: Gilmore 1933:124; H. Smith 1923:61, 1932:353,396.

Sagittaria latifolia Willd.
Common arrowleaf
Native herb; common to occasional throughout Illinois in swamps, sloughs, ponds, and along shorelines.

FOOD: Historic accounts exist of the tubers being eaten by almost all Indian groups in eastern North America.
MEDICINE: Used by the Ojibwa and Potawatomi for poultices and by Ojibwa to aid indigestion.
REFERENCES: Densmore 1928:319, 1929:292; Gilmore 1919:65; H. Smith 1923:124, 1928:254, 1932:94, 1933:37.

ARACEAE
Arum Family

Acorus calamus L.
Sweet flag
Probably native herb; occasional throughout Illinois in marshes and low areas.

FOOD: Rootstock eaten by the Indians of New York State.
MEDICINE: Rootstock used by numerous Indian groups as a physic, a stimulant, and to treat coughs, colds, fever, hemorrhage, catarrh, burns, and toothache. The plant is suspected of being hallucinogenic, especially in large doses.
TECHNOLOGY: Leaves used for wigwam thatch by the Menominee and as a fish attractant by the Ojibwa.
SMOKING: Smoked by the Ojibwa.
REFERENCES: Densmore 1928:286; Epstein 1981:34; Gilmore 1919:17, 1933:124; Lewis and Elvin-Lewis 1977:273,294,301,306,307,404; H. Smith 1923:22, 1928:201, 1932:355,428, 1933:39; Tehon 1951:15; Vogel 1970:201; Yanovsky 1936:10.

Arisaema dracontium (L.)Schott
Green dragon
Native herb; common throughout Illinois in rich woodlands.

MEDICINE: Used by the Menominee for "female troubles" and by the Ojibwa for sore eyes. This plant is poisonous because of the accumulation of calcium oxalate crystals and causes dermatitis.
CEREMONY: Used by the Menominee in sacred bundles to give the power of the supernatural to dreams.
REFERENCES: Lewis and Elvin-Lewis 1977:58,80; H. Smith 1923:23,79, 1932:356; Vogel 1970:321-22.

Arisaema triphyllum (L.)Schott
Jack-in-the-pulpit, Indian turnip
Native herb; common throughout Illinois in rich woods.

FOOD: Corm eaten by the Menominee.
MEDICINE: Used by the Meskwaki to reduce swelling of snakebites and by Menominee and Ojibwa as a poultice for sore eyes; powdered root used by the Pawnee to treat headache. Recent--used as a stimulant, expectorant, irritant, and diaphoretic. The plant is poisonous because of accumulated calcium oxalate crystals and causes dermatitis.
REFERENCES: Densmore 1928:287; Gilmore 1919:17; Lewis and Elvin-Lewis 1977:58,80,169; H. Smith 1923:23, 1928:202, 1933:95; Tehon 1951:25; Vogel 1970:321-22; Yanovsky 1936:10.

Peltandra virginica (L.)Kunth
Arrow arum
Native herb; occasional in the southern four-fifths of Illinois in muddy ponds, shallow water, and swamps.

FOOD: Roots eaten by Indians in the East after lengthy cooking to destroy the calcium oxalate crystals.
REFERENCES: Yanovsky 1936:11.

Symplocarpus foetidus (L.)Nutt.
Skunk cabbage
Native herb; occasional in the northern three-fifths of Illinois in swamps and low areas.

FOOD: Rootstock used as emergency food by the Iroquois and young leaves and shoots cooked and eaten as "greens." This plant is poisonous because of the accumulation of calcium oxalate crystals.
MEDICINE: Root used by the Ojibwa and Menominee for poultices and by the Ojibwa as cough medicine; odor of the crushed leaves inhaled by the Micmacs to treat headache. Recent--used as an antispasmodic.
REFERENCES: Gilmore 1933:124; Lewis and Elvin-Lewis 1977:58,169,301; H. Smith 1923:23; Tehon 1951:110; Yanovsky 1936:11.

COMMELINACEAE
Spiderwort Family

Tradescantia ohiensis Raf.
Spiderwort
Native herb; common throughout Illinois along the edges of woods and in prairies.

SYNONYMS: *T. reflexa* Raf.
MEDICINE: Used by the Meskwaki to treat insanity.
REFERENCES: H. Smith 1928:198.

Tradescantia virginiana L.
Spiderwort
Native herb; common in woods and prairies of the southern two-thirds of Illinois, rare elsewhere.

CEREMONY: Used by the Indians of the Missouri River region as a love charm.
REFERENCES: Gilmore 1919:18.

CYPERACEAE
Sedge Family

Carex plantaginea Lam.
Plantain-leaved sedge
Native herb; occurs in woods in the northeastern portion of Illinois; rare.

MEDICINE: Juice of chewed root applied to snakebites by the Menominee.
CEREMONY: Plant used as a charm against rattlesnakes by the Menominee.
REFERENCES: H. Smith 1923:34.

Cyperus aristatus Rotb.
Sedge
Native herb; scattered throughout Illinois on wet, sandy soil.

SYNONYMS: *C. inflexus* Muhl.
FOOD: Tubers eaten in New Mexico. (This is probably a misidentification; this is a very small annual species and Illinois specimens have no indication of tubers.)
REFERENCES: Yanovsky 1936:9.

Cyperus esculentus L.
Nut-sedge
Native herb; common throughout Illinois on moist ground.

FOOD: Tubers eaten in the Southeast.
REFERENCES: Yanovsky 1936:9.

Scirpus acutus Muhl.
Hard-stem bulrush
Native herb; occasional throughout Illinois in shallow water.

FOOD: Roots, seeds, and young shoots eaten in eastern North America.
REFERENCES: Yanovsky 1936:10.

Scirpus cyperinus (L.)Kunth
Wool-grass
Native herb; common throughout Illinois on wet ground.

TECHNOLOGY: Stems used for weaving mats and bags by the Ojibwa; fruiting tops used for stuffing pillows by the Potawatomi.
REFERENCES: H. Smith 1932:418, 1933:118.

Scirpus validus Vahl
Soft-stem Bulrush
Native herb; common throughout Illinois in marshes and shallow water.

SYNONYMS: *S. lacustris* Mead.
FOOD: Root eaten raw or pounded into flour by eastern Indians; pollen sometimes eaten.
TECHNOLOGY: Stem used for weaving mats and baskets by the Ojibwa, Menominee, Potawatomi, and Meskwaki.
REFERENCES: Densmore 1928:292,329; Gilmore 1919:69, 1933:124; H. Smith 1923:74, 1928:268, 1932:418, 1933:118; Yanovsky 1936:10.

DIOSCOREACEAE
Yam Family

Dioscorea villosa L.
Wild yam
Native herb; common throughout Illinois in dry or moist woods.

MEDICINE: Root used by the Meskwaki to relieve pain during childbirth; possibly used by the Menominee as well. Recent--root considered useful as a uterine sedative and hemostatic; the genus is teratogenic and a source of corticosteriods.
REFERENCES: Lewis and Elvin-Lewis 1977:93,95,322; H. Smith 1923:34, 1928:220; Tehon 1951:49; Vogel 1970:234,294.

IRIDACEAE
Iris Family

Iris shrevei Small
Wild blue iris
Native herb; common in wet situations throughout Illinois.

SYNONYMS: *I. virginica* L. var. *shrevei* (Small) E. Anderson; *I. versicolor* L. is closely related.
MEDICINE: Roots of *I. versicolor* were used by the Meskwaki for colds and lung trouble, by the Potawatomi and Ojibwa as poultices, and by the Indians of the Missouri River region for earache. Recent--the dried rhizome has been used as a cathartic, emetic, diuretic, alterative, and purgative. This plant has caused livestock poisoning.
TECHNOLOGY: Leaves used by the Potawatomi for weaving mats and baskets.
CEREMONY: Used as a charm against snakes by the Ojibwa.
REFERENCES: Densmore 1928:290; Gilmore 1919:20, 1933:120; Lewis and Elvin-Lewis 1977:62,344; H. Smith 1928:224, 1932:371,430, 1933:60,120; Vogel 1970:283-84.

Sisyrinchium campestre Bickn.
Blue-eyed grass
Native herb; occurs in prairies, especially in sandy soil; generally restricted to northwest and west-central Illinois.

SYNONYMS: *S. albidum* Raf. is a similar species.
MEDICINE: Water in which plant has been boiled was used to treat hay fever by the Meskwaki.
REFERENCES: H. Smith 1928:224.

JUNCACEAE
Rush Family

Juncus dudleyi Wieg.
Rush
Native herb; occasional in ditches and on marshy ground throughout Illinois.

TECHNOLOGY: Stems used by the Ojibwa for weaving bags, pouches, and mats.
REFERENCES: Gilmore 1933:125.

Juncus effusus L.
Soft rush
Native herb; common in meadows in the southern one-half of Illinois; occasional elsewhere.

TECHNOLOGY: Stems used for weaving small, fine mats by the Ojibwa.
REFERENCES: H. Smith 1932:419.

JUNCAGINACEAE
Arrow-grass Family

Triglochin maritima L.
Arrow-grass
Native herb; occurs locally in sandy shores, swamps, and wet ditches in the northeastern portion of Illinois.

FOOD: Seeds eaten by Indians in the western United States.
REFERENCES: Yanovsky 1936:7.

LILIACEAE
Lily Family

Aletris farinosa L.
Colic-root
Native herb; occurs in moist sandy prairies and flats in the northern one-third of Illinois.

MEDICINE: Used by Indians in the Southeast to treat fevers, sore breasts, colic, and dysentery and as a tonic, stomachic, and narcotic. Recent--used as uterine tonic, diuretic, and treatment for digestive troubles.
REFERENCES: Tehon 1951:17; Vogel 1970:375-76.

Allium canadense L.
Wild garlic
Native herb; common throughout Illinois in dry woods, prairies, and waste ground.

FOOD: Bulbs used for food and seasoning by the Menominee, Potawatomi, Iroquois, and Sauk-Fox.
REFERENCES: Parker 1910:105; H. Smith 1923:69, 1928:262, 1933:104; Waugh 1916:118; Yanovsky 1936:11.

Allium cernuum Roth
Nodding onion
Native herb; occurs on banks in the northeastern portion of Illinois; uncommon.

FOOD: Bulbs cooked for food by the Ojibwa and Menominee.
REFERENCES: H. Smith 1932:406; Yanovsky 1936:11.

Allium mutabile Michx.
Wild onion
Native herb; occurs in dry areas, uncommon.

FOOD: Bulbs used by the Indians of the Missouri River region for food.
REFERENCES: Gilmore 1919:71; Yanovsky 1936:11.

Allium stellatum Fraser
Wild onion
Native herb; occurs in hill prairies in the southwestern portion of Illinois; rare.

FOOD: Bulbs used for food in British Columbia.
MEDICINE: Bulbs used by the Ojibwa to treat colds.
REFERENCES: Densmore 1928:286; Yanovsky 1936:11.

Allium trioccum Ait.
Wild leek
Native herb; occasional in moist, rich woods in the northern one-half of Illinois; rare elsewhere.

SYNONYMS: *A. burdickii* (Hanes) A. G. Jones is a similar species.
FOOD: Bulbs eaten raw or dried as food or seasoning by the Ojibwa, Menominee, Meskwaki, and Potawatomi.
MEDICINE: Stalk decoction used by the Ojibwa as an emetic.
REFERENCES: Densmore 1928:286; Parker 1910:105; H. Smith 1923:69, 1928:262, 1932:406, 1933:104; Vogel 1970:306-307; Waugh 1916:118; Yanovsky 1936:112.

Chamaelirium luteum (L.)Gray
Fairy wand
Native herb; occurs on low wooded hillsides in Hardin, Massac, and Pope counties; rare.

MEDICINE: Recent--the rhizome has been used as a uterine tonic, diuretic, and anthelminthic.
REFERENCES: Tehon 1951:36; Vogel 1970:282.

Convallaria majalis L.
Lily-of-the-valley
Introduced herb; cultivated and occasionally escaped.

MEDICINE: Recent--the rootstock has been used as a cardiac tonic and shows experimental hypoglycemic effect. The plant is poisonous.
DYE: Recent--the young leaves can be used for yellow to greenish-yellow; autumn leaves for gold.
REFERENCES: Brooklyn Botanic Garden 1964:24; Lust 1974:557; Tehon 1951:43.

Erythronium albidum Nutt.
White dog-tooth violet
Native herb; common throughout Illinois on alluvial soil in woods.

FOOD: Small bulbs eaten by Indian children in the Missouri River region. Early spring bulbs have caused poisoning in poultry.
REFERENCES: Gilmore 1919:71; Lewis and Elvin-Lewis 1977:60; Yanovsky 1936:13.

Erythronium americanum Ker
Yellow dog-tooth violet
Native herb; occasional in moist woods in the northeastern portion of Illinois.

FOOD: Bulbs eaten by Winnebago children. Early spring bulbs have caused poisoning in poultry.
MEDICINE: Herb and bulb have been used to treat gout.
REFERENCES: Lewis and Elvin-Lewis 1977:60; Tehon 1951:53; Vogel 1970:269.

Lilium philadelphicum L. var. *andinum* (Nutt.) Ker
Wood lily
Native herb; occasional in dry open woods and prairies in the northern one-half of Illinois.

MEDICINE: Used by the Indians of the Missouri River region as a treatment for snakebite; bulbs used by the Ojibwa as a poultice for wounds and snakebite.
REFERENCES: Gilmore 1919:19, 1933:125.

Maianthemum canadense Desf.
Wild lily-of-the-valley
Native herb; occasional in moist woods in the northern one-fourth of Illinois.

FOOD: Berries eaten by the Potawatomi.
MEDICINE: Root used by the Potawatomi to treat sore throats and by Ojibwa to clear kidneys during pregnancy.
CEREMONY: Root used by the Potawatomi as a hunting charm.
REFERENCES: H. Smith 1932:374, 1933:62,105,121; Yanovsky 1936:14.

Medeola virginiana L.
Indian cucumber-root
Native herb; occurs in wooded ravines in northeastern Illinois; rare.

FOOD: Tubers eaten by Indians in the Northeast.
REFERENCES: Yanovsky 1936:14.

Polygonatum biflorum (Walt.)Ell.
Small Solomon's seal
Native herb; occasional in dry woods and on sandstone cliffs in the southern tip of Illinois.

MEDICINE: Fumes of burning root used by the Meskwaki and Menominee to revive an unconscious person; root tea used by the Ojibwa to treat coughs and as a physic. Recent--mucilaginous root used as a demulcent. The berries are known to have caused vomiting and diarrhea.
REFERENCES: Lewis and Elvin-Lewis 1977:61; H. Smith 1928:230, 1932:374; Tehon 1951:89.

Polygonatum commutatum (Schult.)A.Dietr.
Solomon's seal
Native herb; common throughout Illinois in woods.

FOOD: Rootstock eaten by the Iroquois.
MEDICINE: Fumes of burning root inhaled by Ojibwa for headache; also used for measles. Recent--mucilaginous root used as a demulcent.
REFERENCES: Densmore 1928:291; Gilmore 1933:125; Tehon 1951:89; Yanovsky 1936:14.

Smilacina racemosa (L.)Desf.
False Solomon's seal
Native herb; common throughout Illinois in rich, moist woods.

SYNONYMS: *Vagnera racemosa* (L.)Morong.
FOOD: Berries eaten by the Ojibwa and in the Northeast.
MEDICINE: Root used by the Ojibwa to treat lung troubles, "female troubles," and headaches; smudged by the Meskwaki for severe illnesses, used by Potawatomi as a reviver, and by Menominee for catarrh.
REFERENCES: Densmore 1928:294; Gilmore 1933:125; H. Smith 1923:41, 1928:230, 1932:374, 1933:63; Vogel 1970:374-75; Yanovsky 1936:15.

Smilacina stellata (L.)Desf.
Small false Solomon's seal
Native herb; occasional in moist woods and prairies in the northern three-fifths of Illinois.

FOOD: Berries eaten by Indians in British Columbia.
MEDICINE: Root infusion used to regulate menstrual disorders and leaf tea used to prevent conception by Nevada Indians.
REFERENCES: Vogel 1970:244; Yanovsky 1936:15.

Trillium cernuum L. var. *macranthum* Eames & Wieg.
Nodding trillium
Native herb; occurs in moist woodlands in Cook and McHenry counties; rare.

MEDICINE: Recent--root used as an astringent, tonic, alterative, and emetic.
REFERENCES: Vogel 1970:66.

Trillium grandiflorum (Michx.)Salisb.
Large white trillium
Native herb; occasional in rich, moist woods in the northern one-half of Illinois; rare elsewhere.

FOOD: The greens are supposedly edible although there is no evidence for historic or aboriginal usage.
MEDICINE: A root infusion was used by the Potawatomi to treat sore nipples, by the Menominee for sore eyes, and by the Ojibwa for sore ears, cramps, and rheumatism. A root tea was used by some North American Indians to facilitate parturition and regularize menstruation. Recent--used as an astringent, tonic, alterative, and emetic.
REFERENCES: Densmore 1928:293; Lewis and Elvin-Lewis 1977:322; H. Smith 1923:41, 1933:63; Vogel 1970:385.

Uvularia grandiflora Sm.
Yellow Bellwort
Native herb; common throughout Illinois in rich woodlands.

MEDICINE: Root used by the Ojibwa to treat stomach trouble, by the Potawatomi for headaches and sore muscles, and by the Menominee to reduce swellings.
REFERENCES: H. Smith 1923:41, 1932:374, 1933:64.

Uvularia sessifolia L.
Bellwort
Native herb; occurs in woods in the southern one-third of Illinois; rare.

FOOD: Young shoots eaten like asparagus by Indians in the eastern states.
CEREMONY: Used as part of a hunting charm by the Ojibwa.
REFERENCES: H. Smith 1932:430, 1933:15.

ORCHIDACEAE
Orchid Family

Aplectrum hyemale (Muhl.)Torr.
Putty-root orchid
Native herb; occasional throughout Illinois in rich woods.

MEDICINE: Recent--corm contains mucilage which has been considered useful as a demulcent and pectoral.
REFERENCES: Tehon 1951:21.

Corallorhiza odontorhiza (Willd.)Nutt.
Coral-root orchid
Native herb; occurs in woodlands throughout Illinois; uncommon.

MEDICINE: Recent--rootstock has been used as a diaphoretic.
REFERENCES: Tehon 1951:43.

Cypripedium acaule Ait.
Lady's-slipper
Native herb; occurs in bogs and acid woodlands in the northeastern portion of Illinois; rare.

MEDICINE: Root used by the Menominee for "male troubles" and by the Meskwaki as a love medicine. Recent--used as a nerve stimulant and antispasmodic.
CEREMONY: Used by the Meskwaki as a love medicine.
REFERENCES: H. Smith 1923:44, 1928:234; Vogel 1970:330.

Cypripedium calceolus L. var. *pubescens* (Willd.)Correll
Yellow lady's-slipper
Native herb; occasional throughout Illinois in moist or dry woodlands.

SYNONYMS: *C. parviflorum* Salisb.
MEDICINE: Root used for "female troubles" by the Ojibwa and Menominee. Recent--used as a mild nerve stimulant and antispasmodic.
CEREMONY: Placed by the Menominee in sacred bundles to induce dreams.
REFERENCES: H. Smith 1923:44, 1932:377; Tehon 1951:46; Vogel 1970:330.

Cypripedium reginae Walt.
Showy lady's-slipper
Native herb; occurs in bogs and low, wet areas in the northern one-half of Illinois; rare.

MEDICINE: Used for "medicine" by the Ojibwa. Recent--used as a nerve stimulant and antispasmodic.
REFERENCES: Densmore 1928:289; Vogel 1970:330.

Habenaria viridis (L.)R.Br. var. *bracteata* (Muhl.)Gray
Bracted green orchid
Native herb; occurs in rich woodlands in the northern one-half of Illinois; rare.

SYNONYMS: *H. bracteata* (Muhl.)R.Br.
MEDICINE: Smuggled into food by the Ojibwa as an aphrodisiac.
REFERENCES: H. Smith 1932:377,431.

Malaxis unifolia Michx.
Adder's-mouth orchid
Native herb; occurs in dry or moist woodlands; rare.

SYNONYMS: *Mycrostylis unifolia* (Michx.)B.S.P.
MEDICINE: Root used as a diuretic by the Ojibwa.
REFERENCES: H. Smith 1932:377.

Spiranthes gracilis (Bigel.)Beck
Slender ladies-tresses
Native herb; occurs in open woods in the southern one-third of Illinois; rare.

CEREMONY: Root used as an ingredient in an Ojibwa hunting charm.
REFERENCES: H. Smith 1932:431.

POACEAE
Grass Family

Agropyron repens (L.)Beauv.
Quack grass
Introduced grass; common in fields and waste ground in the northern three-fourths of Illinois; rare elsewhere.

MEDICINE: Recent--the rootstock sometimes used as a demulcent and diuretic.
REFERENCES: Tehon 1951:16.

Andropogon gerardii Vitman
Big bluestem
Native grass; occasional to common in prairies throughout Illinois.

SYNONYMS: *A. furcatus* Muhl.
MEDICINE: Used by the Ojibwa for indigestion.
TECHNOLOGY: Used by the Omaha in earthen lodge construction.
REFERENCES: Densmore 1928:286; Gilmore 1919:68.

Andropogon virginicus L.
Broom-sedge
Native grass; occasional to common in fields and open woods in the southern one-half of Illinois.

DYE: Recent--the stalks and leaves can be used for greenish and yellowish colors.
REFERENCES: Brooklyn Botanic Garden 1964:24, 1973:20; Lust 1974:551.

Anthoxanthum odoratum L.
Sweet vernal grass
Introduced grass; occurs in meadows and waste places in the northeastern portion of Illinois; rare.

MEDICINE: Recent--source of an anticoagulant.
TECHNOLOGY: Used by the Potawatomi and Ojibwa in basketry and for sewing bucksins. (The species used was probably *Hierochloe odorata*.)
REFERENCES: Lewis and Elvin-Lewis 1977:192; H. Smith 1932:419, 1933:120.

Arundinaria gigantea (Walt.)Muhl.
Giant cane
Native grass; occasional on river banks and in swamps in the southern one-third of Illinois.

FOOD: Seeds eaten by Indians of the Southeast.
TECHNOLOGY: Stems used by Indians in the Southeast for baskets, knives, arrow shafts, blowguns, fish traps, fences, rafts, and other items.
REFERENCES: Yanovsky 1936:7; Swanton 1946:561,563,565,572,584,592,597.

Avena fatua L.
Wild oat
Introduced grass; occasional in Illinois in fields and waste ground.

FOOD: Seeds eaten by Indians in California.
REFERENCES: Yanovsky 1936:7.

Cinna latifolia (Trevir.)Griseb.
Drooping wood reed
Native grass; occurs in moist woods and along streams in the northeastern portion of Illinois; rare.

SYNONYMS: *C. arundinacea* L. is a similar species.
FOOD: Seeds eaten by western Indians.
REFERENCES: Yanovsky 1936:7.

Elymus arenarius L.
Wild rye
Introduced grass; occurs on sand dunes along Lake Michigan; rare.

SYNONYMS: *E. mollis* Trin.
FOOD: Seeds eaten by California Indians.
REFERENCES: Yanovsky 1936:8.

Elymus canadensis L.
Nodding wild rye
Native grass; common throughout Illinois on roadsides and in dry prairie.

FOOD: Seeds eaten by western Indians.
REFERENCES: Doebley 1984; Yanovsky 1936:8.

Glyceria borealis (Nash.)Batchelder
Northern manna grass
Native grass; occurs in shallow water in northeastern Illinois; rare.

SYNONYMS: *Panicularia borealis* Nash.
FOOD: Seeds eaten by western Indians.
REFERENCES: Doebley 1984; Yanovsky 1936:8.

Glyceria canadensis (Michx.)Trin.
Rattlesnake manna grass
Native grass, occurs on wet ground in northeastern Illinois; rare.

MEDICINE: Root used for "medicine" by the Ojibwa.
REFERENCES: H. Smith 1932:371.

Glyceria striata (Lam.)Hitch.
Fowl manna grass
Native grass; common throughout Illinois on moist ground.

SYNONYMS: *Panicularia nervata* (Willd.)Kuntze
FOOD: Seeds eaten by western Indians. The grass can be poisonous to livestock.
REFERENCES: Doebley 1984; Yanovsky 1936:8.

Hierochloe odorata (L.)Beauv.
Sweet grass
Native grass; occasional in meadows in the northern one-fifth of Illinois.

TECHNOLOGY: Long, sweet-scented leaves used for sewing and weaving bags and baskets by the Ojibwa, Potawatomi, Menominee, Ottawa, and probably others.
CEREMONY: Used ceremonially by the Ojibwa.
SMOKING: Smoked ceremonially by the Crow.
REFERENCES: Densmore 1928:294; Epstein 1981:130; H. Smith 1923:75.

Hordeum jubatum L.
Squirrel-tail grass
Native grass; common in fields and along roadsides in the northern one-half of Illinois; occasional elsewhere.

FOOD: Seeds eaten by western Indians. Bristles may cause mechanical injury or death of livestock.
REFERENCES: Densmore 1928:290; Lewis and Elvin-Lewis 1977:59; H. Smith 1933:59; Yanovsky 1936:8.

Hordeum pusillum Nutt.
Small wild barley
Native grass; occasional to common in fields and along roadsides throughout Illinois.

FOOD: Seeds eaten by southwestern Indians; carbonized seeds found in numerous Illinois archaeological sites.
REFERENCES: Ford 1981; F. King 1983; Yarnell 1983.

Hordeum vulgare L.
Common barley
Introduced grass; occasional throughout Illinois along roadsides.

FOOD: Common cultivated crop.
MEDICINE: Recent--grain sometimes used to prepare barley water, a demulcent.
REFERENCES: Tehon 1951:65.

Koeleria macrantha (Ledeb.)Spreng.
June grass
Native grass; occasional throughout Illinois in sand prairies and black oak woods.

SYNONYMS: *K. cristata* (L.)Pers.
FOOD: Seeds eaten by western Indians.
REFERENCES: Doebley 1984; Yanovsky 1936:8.

Muhlenbergia asperifolia (Nees & Meyer)Paodi
Scratch grass
Native grass; scattered on sandy soil in the northern two-thirds of Illinois.

SYNONYMS: *Sporobolus asperifolius* (Nees & Meyer)Nees
FOOD: Seeds eaten by western Indians.
REFERENCES: Doebley 1984; Yanovsky 1936:9.

Panicum capillare L.
Witch grass
Native grass; common throughout Illinois on waste ground.

FOOD: Seeds eaten by western Indians.
REFERENCES: Doebley 1984; Yanovsky 1936:8.

Phragmites australis Trin.
Reed grass

Native grass; occasional on moist ground in the northern one-half of Illinois; uncommon elsewhere.

SYNONYMS: *Arundo phragmites* L., *P. communis* Trin.
FOOD: Seeds eaten by western Indians.
TECHNOLOGY: Plant used for "utility" by the Ojibwa.
REFERENCES: Densmore 1928:291; Yanovsky 1936:8.

Schizachyrium scoparium (Michx.)Nash
Little bluestem
Native grass; occasional to common throughout Illinois in prairies, fields, and open woods.

SYNONYMS: *Andropogon scoparius* Michx.
MEDICINE: Stem ash used by the Comanches to treat syphilitic sores.
REFERENCES: Vogel 1970:213.

Spartina pectinata Lind.
Cord grass
Native grass; occasional throughout Illinois in wet prairies and marshes.

SYNONYMS: *S. michauxiana* Hitch.
TECHNOLOGY: Used by the Indians of the Missouri River region for thatching.
REFERENCES: Gilmore 1919:14.

Sporobolus cryptandrus (Torr.)Gray
Sand dropseed
Native grass; occasional in the northern one-third of Illinois on sandy soil.

FOOD: Seeds eaten by western Indians.
REFERENCES: Doebley 1984; Yanovsky 1936:9.

Stipa spartea Trin.
Porcupine grass
Native grass; occasional on sandy soil in the northern two-thirds of Illinois.

TECHNOLOGY: Long-awned caryopses used by the Indians of the Missouri River region as brushes. Awns can cause mechanical injury and death to livestock.
REFERENCES: Gilmore 1919:14.

Vulpia octoflora (Walt.)Rydb.
Six-weeks fescue
Native grass; occasional throughout Illinois on sandy soil.

SYNONYMS: *Festuca octoflora* Walt.
FOOD: Seeds eaten by western Indians.
REFERENCES: Doebley 1984; Yanovsky 1936:8.

Zizania aquatica L.
Wild rice
Native grass; occurs on streambanks and in shallow water in ponds, lakes, streams, and marshes in the northern three-fourths of Illinois; uncommon.

FOOD: One of the most important cereals among North American Indians; used throughout its range.
MEDICINE: Gruel served to convalescents by numerous Indian groups.
REFERENCES: Gilmore 1919:67; H. Smith 1923:67, 1928:259; Vogel 1970:200; Yanovsky 1936:9.

SMILACACEAE
Catbrier Family

Smilax bona-nox L.
Catbrier
Native vine; occasional in dry woods and fields and on bluffs, in the southern one-fourth of Illinois.

FOOD: Tuberous rootstocks eaten by Indians in the Southeast.
MEDICINE: Recent--the rootstock has been used as a substitute for true sarsaparilla and as an alterative and a diuretic.
REFERENCES: Tehon 1951:106; Vogel 1970:51; Yanovsky 1936:14.

Smilax glauca Walt.
Catbrier
Native vine; common in dry woods, on bluffs and along fields in the southern one-third of Illinois.

FOOD: Rootstock eaten by Indians of the Southeast.
MEDICINE: Roots used by Alabama Indians to treat stomach trouble and injured backs.
REFERENCES: Vogel 1970:235,366; Yanovsky 1936:14.

Smilax herbacea L.
Carrion flower
Native herb; occurs in moist woods in Jackson County; rare.

SYNONYMS: *S. lasioneuron* Hook. is a similar species.
MEDICINE: Root used by the Ojibwa to treat lung trouble.
REFERENCES: H. Smith 1932:374.

Smilax hispida Muhl.
Bristly catbrier
Native vine; occasional throughout Illinois in woods and moist thickets.

CEREMONY: Used as "bad magic" by the Ojibwa.
REFERENCES: Gilmore 1933:126.

Smilax rotundifolia L.
Catbrier
Native vine; common in dry woods in the southern one-third of Illinois.

FOOD: Tuberous rootstocks eaten by Indians in the Southeast.
REFERENCES: Yanovsky 1936:14.

SPARGANIACEAE
Bur-reed Family

Sparganium eurycarpum Engelm.
Bur-reed
Native herb; occasional throughout Illinois in shallow water.

FOOD: Tubers and stem bases eaten by northwestern Indians.
REFERENCES: Yanovsky 1936:7.

TYPHACEAE
Cat-tail Family

Typha angustifolia L.
Narrow-leaved cat-tail
Native herb; occasional throughout Illinois in wet ground.

FOOD: Young roots, shoots, stems, flowers, and seeds eaten by western Indians.
REFERENCES: Yanovsky 1936:6.

Typha latifolia L.
Common cat-tail
Native herb; common throughout Illinois in marshes, ditches, and pond margins.

FOOD: Roots, shoots, stems, flowers, and seeds eaten by western Indians.
TECHNOLOGY: Leaves used by the Menominee, Ojibwa, and Potawatomi to make mats to thatch wigwams; roots used by the Menominee to repair boats; fuzz used by the Indians of the Missouri River region to pad cradleboards.
CEREMONY: Fuzz thrown by the Ojibwa into the faces of the enemy to blind them.
REFERENCES: Densmore 1928:293; Gilmore 1919:12, 1933:124; H. Smith 1923:77, 1932:423,432, 1933:114; Yanovsky 1936:6.

SUBDIVISION II: ANGIOSPERMAE
CLASS II: DICOTYLEDONEAE

Plants usually characterized by an embryo with two cotyledons, leaves that are net- rather than parallel-veined, and flowers with parts in fours or fives (e.g., four petals and four sepals).

ACERACEAE
Maple Family

Acer negundo L.
Box elder
Native tree; common throughout Illinois on moist soil.

FOOD: Sap used to make sugar and as a seasoning by the Ojibwa.
MEDICINE: Used by the Ojibwa and Meskwaki as an emetic. Recent--has shown some potential in cancer chemotherapy.
REFERENCES: Gilmore 1919:101; Lewis and Elvin-Lewis 1977:133; H. Smith 1928:200, 1932:353,394; Yanovsky 1936:4.

Acer rubrum L.
Red maple
Native tree; occasional in the southern third of Illinois in rocky or moist woods and on slopes.

FOOD: Sap used by the Iroquois to make sugar and the bark pounded for bread. Recent--the leaves have caused poisoning in livestock.
MEDICINE: Cambium used by the Ojibwa and Potawatomi to make eyewash.
TECHNOLOGY: Used by the Potawatomi to deodorize traps and by the Ojibwa in beadwork designs.
DYE: Recent--bark used to make olive-gray.
REFERENCES: Lewis and Elvin-Lewis 1977:48; Lust 1974:560; H. Smith 1932:353,412, 1933:37,116; Waugh 1916:119.

Acer saccharum Marsh.
Sugar maple
Native tree; common throughout Illinois in rich woods.

FOOD: Sap used for sugar and seasoning and cambium pounded to make bark bread by virtually all tribes in the geographical range of the species.
MEDICINE: Inner bark used by the Meskwaki as an emetic, by the Potawatomi as an expectorant, and by the Ojibwa for "medicine."

TECHNOLOGY: Wood used by Ojibwa for bowls, food stirring-paddles and arrows.
DYE: Used by the Indians of the Missouri River region for black.
REFERENCES: Densmore 1928:286; Gilmore 1919:48,100; H. Smith 1923:61, 1928:196,255, 1932:394,413, 1933:37,92; Waugh 1916:119.

Acer saccharinum L.
Silver maple
Native tree; common throughout Illinois in bottomland woods and along streams.

FOOD: Sap used for sugar and seasoning and cambium pounded to make bark bread by Ojibwa and Menominee and undoubtedly others.
REFERENCES: Gilmore 1919:100; H. Smith 1933:136; Waugh 1916:119.

AMARANTHACEAE
Amaranth Family

Amaranthus hybridus L.
Green amaranth
Native herb; common throughout Illinois in fields and on waste ground.

FOOD: Young plants eaten as "greens" by Indians in the West.
REFERENCES: Yanovsky 1936:23.

Amaranthus retroflexus L.
Rough pigweed
Native herb; common throughout Illinois in fields and on waste ground.

FOOD: Seeds eaten in the west; leaves eaten as "greens" by the Iroquois. Note--this plant accumulates nitrates from the soil and has caused nitrate poisoning in livestock.
MEDICINE: Recent--herb and root used as astringent and detergent; pollen used as antigen for hay fever
REFERENCES: Lewis and Elvin-Lewis 1977:33; Tehon 1951:18; Waugh 1916:117; Yanovsky 1936:23.

ANACARDIACEAE
Cashew Family

Rhus aromatica Ait.
Fragrant sumac
Native shrub; occasional throughout Illinois in woods and on bluffs and dunes.

SYNONYMS: *R. canadensis* Marsh.; *R. trilobata* Nutt. var. *arenaria* (Greene)Barkley is a similar species.
MEDICINE: Aromatic bark used by Comanches for treating colds; root bark used elsewhere to treat bladder and kidney problems.
SMOKING: Leaves smoked by the Menominee.
REFERENCES: Epstein 1981:31; Lewis and Elvin-Lewis 1977:315; Tehon 1951:95; Vogel 1970:377.

Rhus copallina L.
Dwarf sumac
Native shrub; common in woodlands and fields in the southern one-fourth of Illinois, occasional to rare elsewhere.

FOOD: Fruit used to make a lemonadelike beverage by Indians in the eastern States.
MEDICINE: Root infusion used by the Creeks for dysentery; mixed with tobacco for cephalic and pectoral problems. Recent--leaves, fruit, and bark used as an astringent, refrigerant and gargle.
REFERENCES: Tehon 1951:96; Vogel 1970:377; Yanovsky 1936:40.

Rhus glabra L.
Smooth sumac
Native shrub; common throughout Illinois in woods, fields and along roadsides.

FOOD: Fruit eaten by Indians in the eastern states and used to make a lemonadelike beverage.
MEDICINE: Used by the Ojibwa to treat dysentery and asthma and by the Indians of the Missouri River region as a styptic. Recent--used as an astringent, refrigerant and gargle; root bark used to treat sore mouths and other mucous membrane infections; a 19th century hemorrhoid treatment.
DYE: Roots used for yellow by the Meskwaki, orange and red by the Ojibwa. Recent--fruit, leaves, and roots used for yellow, gray, brown, and orange.
SMOKING: Leaves smoked by the Menominee, Ojibwa, and the Indians of the Missouri River region.
REFERENCES: Brooklyn Botanic Garden 1964:26,27,29, 1973:22; Densmore 1928:291; Gilmore 1919:47, 1933:135; Lewis and Elvin-Lewis 1977:238, 244, 264, 292; Lust 1974:562; H. Smith 1928:271, 1932:354, 424; Tehon 1951:96; Vogel 1970:376-78; Yanovsky 1936:40.

Rhus typhina L.
Staghorn sumac
Native shrub; occasional in woods in the northern half of Illinois.

FOOD: Fruit used to make a lemonadelike beverage by Indians in the northeast; fruit sometimes dried for the winter.
MEDICINE: Root, bark, cambium and berries used by the Meskwaki for pinworms and by the Ojibwa for hemorrhage and stomach pains.
DYE: Used by Ojibwa and Menominee to produce yellow. Recent--fruits and leaves used for gray and black.
SMOKING: Leaves mixed with tobacco by the Potawatomi.
REFERENCES: Brooklyn Botanic Garden 1973:22; Densmore 1928:291; H. Smith 1923:22,77, 1928:201, 1932:354,397,424, 1933:38,95; Tehon 1951:96; Vogel 1970:376-78.

Toxicodendron radicans (L.)Kuntze
Poison ivy
Native vine; common throughout Illinois in fields, woods, and on bluffs and waste ground.

SYNONYMS: *Rhus radicans* L.
MEDICINE: Leaves used by the Potawatomi and Meskwaki in poultice to induce swellings to open. Recent--used to treat cutaneous eruptions.
REFERENCES: H. Smith 1928:201; Tehon 1951:96; Vogel 1970:350.

Toxicodendron vernix (L.)Kuntze
Poison sumac
Native shrub; occasional to rare in the northeastern one-fourth of Illinois.

SYNONYMS: *Rhus vernix* L.
MEDICINE: Recent--used as an antigen for poison ivy.
REFERENCES: Tehon 1951:97.

ANNONACEAE
Custard-Apple Family

Asimina triloba (L.)Dunal
Pawpaw
Native tree; occasional in woods throughout Illinois.

FOOD: Fruit eaten by Iroquois and by Indians of the southeastern and central states.
MEDICINE: A flower tea used by the Seminoles for kidney trouble.
TECHNOLOGY: Bark fiber used by Menominee and Potawatomi for weaving bags.
REFERENCES: Lewis and Elvin-Lewis 1977:315; Waugh 1916:129; Yanovsky 1936:26.

APOCYNACEAE
Dogbane Family

Apocynum androsaemifolium L.
Spreading dogbane
Native herb; common to occasional in woods and prairies throughout Illinois.

MEDICINE: Root used by the Meskwaki to treat snakebite, by the Potawatomi for internal troubles, and by the Ojibwa to cleanse the kidneys and for throat and heart trouble, headache, and colds. Recent--this plant has been found to be poisonous; the root is an emetic in large doses and a tonic in small doses; the sap may cause dermatitis.
TECHNOLOGY: Fiber used for sewing by the Menominee, Potawatomi, and Ojibwa.
CEREMONY: Eaten ceremonially by the Ojibwa and used to keep away evil spirits.
REFERENCES: Densmore 1928:286; Lewis and Elvin-Lewis 1977:51,301,86; H. Smith 1923:73-79, 1928:193, 1932:355,413,428, 1933:38,111; Tehon 1951:22; Vogel 1970:310.

Apocynum cannabinum L.
Indian hemp
Native herb; common throughout Illinois on roadsides and in fields and woods.

MEDICINE: Green fruit used by the Potawatomi for heart and kidney problems and by them and the Meskwaki to treat dropsy. Recent--used in cancer chemotherapy and as a cardiotonic, diuretic, diaphoretic, expectorant, emetic, and cathartic. This plant is poisonous, and the sap may cause dermatitis.
TECHNOLOGY: Fiber used by the Ojibwa to manufacture the best cordage.
REFERENCES: Lewis and Elvin-Lewis 1977:51,86,134,184,193,301,312; H. Smith 1928:201; Tehon 1951:22; Vogel 1970:319.

AQUIFOLIACEAE
Holly Family

Ilex verticillata (L.)Gray
Winterberry
Native shrub; occasional in swamps in the northern one-fourth of Illinois.

FOOD: A beverage made from the leaves by Indians in the Northeast.
MEDICINE: Used by the Ojibwa for diarrhea and by the Meskwaki as a poultice for sores. Recent--leaves, fruit, and bark used as a diaphoretic and demulcent.
REFERENCES: H. Smith 1928:195, 1932:355; Tehon 1951:67; Vogel 1970:270-71; Yanovsky 1936:41.

ARALIACEAE
Ginseng Family

Aralia hispida Vent.
Bristly sarsaparilla
Native herb; occurs on sandy soil in Cook and Lake counties; rare.

MEDICINE: Possibly used as a tonic by the Potawatomi.
REFERENCES: H. Smith 1933:40; Vogel 1970:361.

Aralia nudicaulis L.
Wild sarsaparilla
Native herb; occasional in the northern half of Illinois on rich wooded slopes, dunes, bogs, and in rocky woods.

FOOD: Fruit eaten by Indians in British Columbia.
MEDICINE: Root used by the Meskwaki for "internal troubles" and by the Potawatomi and Ojibwa for poultices, coughs, and sore eyes. Recent--used as an aromatic, stimulant, diaphoretic, and alterative.
TECHNOLOGY: Used as a fish attractant on nets by the Ojibwa.
CEREMONY: Used by the Ojibwa to repel rattlesnakes.
REFERENCES: Densmore 1928:286; H. Smith 1928:195, 1932:356, 1933:40; Tehon 1951:22; Vogel 1970:359-61; Yanovsky 1936:47.

Aralia racemosa L.
American spikenard
Native herb; occasional throughout Illinois in rich or rocky woods.

FOOD: Tips of the young shoots used in soup by the Potawatomi.
MEDICINE: Root tea used by the Cherokees and Appalachian settlers for backache and rheumatoid arthritis. Used as a poultice by the Potawatomi; for coughs by the Ojibwa; root used by the Menominee to treat stomachaches and blood poisoning and as "medicine" by the Meskwaki. Recent--the roots were included in a popular 19th century cough syrup.
REFERENCES: Densmore 1928:287; Lewis and Elvin-Lewis 1977:167; H. Smith 1923:24, 1928:203, 1932:356, 1933:41,96; Tehon 1951:23; Vogel 1970:374-75; Yanovsky 1936:41.

Aralia spinosa L.
Hercules'-club
Native tree; occasional in the southern one-third of Illinois on wooded slopes.

MEDICINE: Supposedly used by "Indians" for fever. Recent--used as a stimulant and diaphoretic.
REFERENCES: Tehon 1951:23; Vogel 1970:273.

Panax quinquefolius L.
Ginseng
Native herb; occasional to rare throughout Illinois in rich or rocky woods.

MEDICINE: Used by the Potawatomi and Menominee as a tonic, by the Meskwaki as a universal medicine, and possibly by the Ojibwa. It was also sold to Euro-Americans; in fact, Indian use may have been stimulated by their demand. Recent--used as an aromatic, mild stimulant and stomachic; the roots are said to stimulate the appetite.
CEREMONY: Used by the Indians of the Missouri River region as a love charm.
REFERENCES: Gilmore 1919:54; Lewis and Elvin-Lewis 1977:372-73,374,273; H. Smith 1923:24, 1928:204, 1932:356, 1933:41; Tehon 1951:84; Vogel 1970:307-10.

ARISTOLOCHIACEAE
Birthwort Family

Aristolochia serpentaria L.
Virginia snakeroot
Native herb; occurs locally in the southern half of Illinois in rich woods.

MEDICINE: Used by several southern tribes to treat snakebite, coughs and fever, to check vomiting, and as a sudorific, stomachic, and tonic. Recent--used as an aromatic and gastric stimulant to treat dyspepsia.
REFERENCES: Lewis and Elvin-Lewis 1977:273,279; Tehon 1951:25; Vogel 1970:373-74.

Asarum canadense L.
Wild ginger
Native herb; common throughout Illinois in rich woods.

SYNONYMS: *A. acuminata* (Ashe)Bickn; *A. reflexum* Bickn.
FOOD: Used by the Menominee, Potawatomi, and Meskwaki as seasoning for food.
MEDICINE: Root used by the Meskwaki for throat trouble, coughing, earache, lung trouble, and stomach cramps and by the Ojibwa for indigestion; used by the Indians of eastern Canada and the Catawbas for heart problems. Recent--used as an aromatic, stimulant, and carminative.
REFERENCES: Densmore 1928:287; Lewis and Elvin-Lewis 1977:193,225,245,360; H. Smith 1928:204; Tehon 1951:26; Vogel 1970:391-92.

ASCLEPIADACEAE
Milkweed Family

Asclepias exaltata L.
Poke milkweed
Native herb; occurs occasionally throughout Illinois in woodland borders.

MEDICINE: Used by the Indians of the Missouri River region to treat stomach trouble. Note-almost all species of *Asclepias* contain cardiac glycosides poisonous to livestock and man; the milky sap may cause dermatitis.
REFERENCES: Gilmore 1919:50; Lewis and Elvin-Lewis 1977:52,86; Vogel 1970:337.

Asclepias incarnata L.
Swamp milkweed
Native herb; common throughout Illinois in fields, prairie edges and waste ground.

FOOD: Flowers, sprouts, buds, and green fruit eaten widely as "greens." See note under *A. exaltata*; this species is suspected of causing poisoning in sheep.
MEDICINE: Used by the Ojibwa and Potawatomi for "female troubles."
TECHNOLOGY: Fiber used for thread and fishlines by the Menominee and Potawatomi.
CEREMONY: Used by the Ojibwa as a deer-hunting charm or attractant.
REFERENCES: Densmore 1928:30; Gilmore 1919:109; Parker 1910:93; H. Smith 1923:62,74, 1928:256, 1932:357,397,428, 1933:96; Tehon 1951:27; Vogel 1970:336-37; Yanovsky 1936:52.

Asclepias syriaca L.
Common milkweed
Native herb; common throughout Illinois in fields and waste ground and along prairie margins.

FOOD: Flowers, sprouts, buds, and green fruit eaten widely by Indians. See note under *A. exaltata* although there is only circumstantial evidence that this species is poisonous.
MEDICINE: Used by the Ojibwa and Potawatomi for "female troubles."
TECHNOLOGY: Fiber used for thread and fishlines by the Menominee and Potawatomi.
CEREMONY: Used by the Ojibwa as a deer-hunting charm or attractrant.
REFERENCES: Densmore:1928:30; Gilmore 1919:109; Parker 1910:93; H. Smith 1923:62,74, 1928:256, 1932:357,397,428, 1933:96; Tehon 1951:27; Vogel 1970:336-37; Yanovsky 1936:52.

Asclepias tuberosa L.
Butterfly-weed
Native herb; common throughout Illinois in woods and prairies.

FOOD: Roots eaten by the Sioux; seedpods, shoots, and buds eaten by eastern Indians. See note under *A. exaltata;* this species is generally considered poisonous.
MEDICINE: Pulverized or macerated roots used by the Menominee for wounds and bruises and by the Missouri River Indians for pulmonary and bronchial troubles. Recent--considered useful as a diuretic, purgative, emetic, expectorant, and diaphoretic.
REFERENCES: Gilmore 1919:57; Lewis and Elvin-Lewis 1977:341; H. Smith 1923:25, 1928:256; Tehon 1951:27; Vogel 1970:287-88; Yanovsky 1936:53.

Asclepias verticillata L.
Horsetail milkweed
Native herb; occurs commonly throughout Illinois in dry rocky woods, prairies, and fields.

MEDICINE: Used by the Euro-American colonists to treat snakebite. This plant is poisonous.
REFERENCES: Vogel 1970:336.

ASTERACEAE
Sunflower Family

Achillea millefolium L. and ssp. *lanulosa* (Nutt.)Piper
Yarrow
Introduced and native herb; common throughout Illinois on roadsides and in fields and lawns.

SYNONYMS: *A. lanulosa* Nutt. is the native element of this species complex.
MEDICINE: Used by the Potawatomi as a reviver; by the Menominee for fevers and eczema; by the Meskwaki for stomach trouble, fever, and body aches; and by the Micmacs for colds. Recent--considered useful as a stimulant, tonic, astringent, and diaphoretic.
CEREMONY: Flowers smoked ceremonially by the Ojibwa; smoke used by the Potawatomi to repel witches.
REFERENCES: Lewis and Elvin-Lewis 1977:306; H. Smith 1923:29, 1928:210, 1933:47,117; Tehon 1951:14.

Ambrosia artemisiifolia L.
Common ragweed
Native herb; common in fields and waste places throughout Illinois.

MEDICINE: Used by the Indians of the Missouri River region for nausea; more recently used as a topical astringent.
REFERENCES: Gilmore 1919:80; Tehon 1951:18.

Ambrosia trifida L.
Giant ragweed

FOOD: Presence in Ozark Bluff shelters has led to the speculation that the seeds may have been used for food; possibly cultivated by Archaic inhabitants of the Lower Illinois Valley.
MEDICINE: Root chewed by the Meskwaki to drive away fear. Recent use same as *A. artemisiifolia*
REFERENCES: Asch and Asch 1982; Gilmore 1931; H. Smith 1928:210; Tehon 1951:19.

Antennaria neglecta Greene
Pussytoes
Native herb; occurs in fields and open woods in the northeastern portion of Illinois, rare.

SYNONYMS: *A. neodioica* Greene
MEDICINE: Used by the Ojibwa to purge afterbirth.
REFERENCES: H. Smith 1932:363.

Anthemis tinctoria L.
Yellow chamomile
Introduced herb; cultivated and occasionally escaped.

DYE: Recent--the flowers can be used to make yellow and gold colors.
REFERENCES: Brooklyn Botanic Garden 1964:22; Lust 1974:554.

Arctium minus (Hill)Bernh.
Common burdock
Introduced herb; occasional to common on waste ground throughout Illinois.

MEDICINE: Used by the Meskwaki to relieve labor pains in women and by the Potawatomi as a general tonic and blood purifier. Root used in 17th century Europe to treat venereal disease. More recently, the root and seeds were considered useful as a diuretic and alterative. The plant can cause contact dermatitis in some persons.
REFERENCES: Lewis and Elvin-Lewis 1977:84,322,333; H. Smith 1928:211, 1933:49; Tehon 1951:24.

Artemisia absinthium L.
Common wormwood
Introduced herb; cultivated and occasionally escaped.

MEDICINE: After its introduction, the plant was used by the Ojibwa for sprains. Recent--leaves and tops considered useful as a stimulant, tonic and a flavoring in certain alcoholic drinks, notably absinthe. The plant can cause contact dermatitis.
REFERENCES: Densmore 1928:287; Lewis and Elvin-Lewis 1977:84; Tehon 1951:26.

Artemisia biennis Willd.
Biennial wormwood
Introduced herb; occasional on waste ground throughout Illinois.

FOOD: Seeds eaten in Utah and Nevada.
REFERENCES: Yanovsky 1936:59.

Artemisia dracunculus L.
False tarragon
Native herb; occasional in prairies in the northern one-half of Illinois.

MEDICINE: Used by the Ojibwa for sprains and by the Missouri River Indians for fevers and rheumatism.
CEREMONY: Used by the Missouri River Indians as both love and hunting charms.
REFERENCES: Densmore 1928:287; Gilmore 1919:82.

Artemisia frigida Willd.
Wormwood
Introduced herb; occurs on waste ground; rare.

MEDICINE: Planted by Potawatomi for use as a reviver.
REFERENCES: H. Smith 1933:49.

Artemisia ludoviciana Nutt. var. *gnaphalodes* (Nutt.)T&G.
White sage
Introduced herb; occasional on waste ground throughout Illinois.

SYNONYMS: *A. gnaphalodes* Nutt.
MEDICINE: Used by the Ojibwa as an "antidote" and by the Arikara for menstrual problems.
CEREMONY: Burned ceremonially by the Cheyenne.
REFERENCES: Densmore 1928:287; Epstein 1981:53; Vogel 1970:238.

Artemisia ludoviciana Nutt.
White sage
Native herb; occasional on waste ground in the northern half of Illinois.

MEDICINE: Used by the Meskwaki for poultices and for sore throats and tonsillitis; by the Ojibwa for "medicine" and taken by Arikara women during menstruation.
CEREMONY: Used as a charm by the Ojibwa.
SMOKING: Smoked by the Sioux.
REFERENCES: H. Smith 1928:211, 1932:363,417; Vogel 1970:238.

Aster cordifolius L.
Blue wood aster
Native herb; occasional in dry woods throughout Illinois.

SMOKING: Root included in a hunting medicine smoked by the Ojibwa to attract deer.
REFERENCES: H. Smith 1932:428.

Aster sagittifolius Wedem. *ex* Willd. var. *drummondii* (Lindl.) Shinners
Drummond's aster
Native herb; common throughout Illinois in dry, open woods.

SYNONYMS: A. *drummondii* Lindl.
MEDICINE: Possibly used for medicine by the Meskwaki.
REFERENCES: H. Smith 1928:211.

Aster ericoides L.
Heath aster
Native herb; occurs throughout Illinois in prairies and dry open ground; occasional.

SYNONYMS: *A. multiflorus* Ait.
MEDICINE: Used by the Meskwaki to revive unconscious persons.
REFERENCES: H. Smith 1928:212.

Aster furcatus Burgess
Forked aster
Native herb; occasional in dry woods in the northern half of Illinois.

MEDICINE: Used by the Potawatomi to treat headaches.
REFERENCES: H. Smith 1933:49.

Aster laevis L.
Smooth aster
Native herb; common throughout Illinois in moist, sandy soil in woods and prairie.

MEDICINE: Used by the Meskwaki as a reviver.
REFERENCES: H. Smith 1928:211.

Aster lateriflorus (L.)Britt.
Side-flowered aster
Native herb; common in woods throughout Illinois.

MEDICINE: Used by the Meskwaki to treat insanity.
REFERENCES: H. Smith 1928:212.

Aster macrophyllus L.
Big-leaved aster
Native herb; occurs in dry, open woods in the northeastern portion of Illinois; rare.

FOOD: Young shoots eaten by the Ojibwa.
MEDICINE: Root used by the Ojibwa to treat headaches and by the Meskwaki to treat a woman whose child was born dead.
SMOKING: Smoked by the Ojibwa to attract deer.
REFERENCES: H. Smith 1928:195, 1932:398,429; Yanovsky 1936:60.

Aster novae-angliae L.
New England aster
Native herb; common throughout Illinois on moist ground.

MEDICINE: Used by the Potawatomi as a fumigant.
SMOKING: Powdered root smoked by the Ojibwa to attract deer.
REFERENCES: Densmore 1928:287; H. Smith 1928:212, 1933:50.

Aster puniceus L.
Swamp aster
Native herb; occasional on moist ground in the northern two-thirds of Illinois, rare elsewhere.

CEREMONY: Used by the Ojibwa as a charm.
REFERENCES: Densmore 1928:287.

Aster shortii Lindl.
Short's aster
Native herb; common in dry woods throughout Illinois.

MEDICINE: Flowers used by the Potawatomi to make a medicinal beverage.
REFERENCES: H. Smith 1933:50.

Aster prealtus Poir.
Willow aster
Native herb; common on moist ground throughout Illinois.

MEDICINE: Used by the Meskwaki as a "reviver" for unconscious persons.
REFERENCES: H. Smith 1928:212.

Aster umbellatus Mill.
Flat-top aster
Native herb; occasional on moist ground in the northern one-half of Illinois.

MEDICINE: Flowers smudged by the Potawatomi to expel evil spirits from the sick.
REFERENCES: H. Smith 1933:50.

Cacalia atriplicifolia L.
Pale Indian-plantain
Native herb; occasional in moist or dry open woods and prairie throughout Illinois.

MEDICINE: Used by the Cherokee for "medicine."
REFERENCES: Vogel 1970:102.

Centaurea cyanus L.
Corn flower
Introduced herb; cultivated and occasionally escaped.

MEDICINE: Tea used in the Mediterranean region as a tonic and stimulant.
DYE: Flowers used to make blue.
REFERENCES: Lewis and Elvin-Lewis 1977:374; Lust 1974:554.

Chrysanthemum leucanthemum L.
Ox-eye daisy
Introduced herb; common in fields, waste places, and along roadsides throughout Illinois.

MEDICINE: Used by the Menominee to treat fevers.
REFERENCES: H. Smith 1923:29.

Cichorium intybus L.
Chicory
Introduced herb; common on waste ground and roadsides throughout Illinois.

FOOD: Present--ground roots used as a bitter substitute or adulterant of coffee; root and greens eaten.
MEDICINE: Used to increase appetite or aid digestion.
REFERENCES: Lewis and Elvin-Lewis 1977:386; Tehon 1951:39.

Cirsium arvense (L.)Scop.
Canada thistle
Introduced herb; occasional in fields and waste ground throughout Illinois.

MEDICINE: Used by the Ojibwa as a bowel tonic and for "female troubles." Recent--considered useful as an antiphlogistic, tonic, and diuretic.
REFERENCES: Densmore 128:208; H. Smith 1932:364; Tehon 1951:40.

Cirsium discolor (Mulh.)Spreng.
Field thistle
Native herb; common throughout Illinois in fields, open woods, and on waste ground.

MEDICINE: Used by the Meswaki to treat stomachache and by the Ojibwa for "female troubles."
REFERENCES: Densmore 1928:288; H. Smith 1928:213.

Coreopsis palmata Nutt.
Prairie coreopsis
Native herb; common in prairies and open woods in the northern one-half of Illinois.

MEDICINE: Made into a medicinal beverage or poultice by the Meskwaki to treat crippled persons.
DYE: Recent--flowers used for yellow.
REFERENCES: Brooklyn Botanic Garden 1973:33; H. Smith 1928:213.

Coreopsis tinctoria Nutt.
Golden coreopsis
Introduced herb; scattered on waste ground throughout Illinois.

DYE: Recent--flowers used for orange-red and yellow.
REFERENCES: Brooklyn Botanic Garden 1964:31, 1973:33; Lust 1974:553.

Coreopsis tripteris L.
Tall tickseed
Native herb; common in open woods and prairies throughout Illinois.

MEDICINE: Made into a medicinal beverage used by the Meskwaki for internal pain or bleeding.
REFERENCES: H. Smith 1928:213.

Dyssodia papposa (Vent.)Hitchc.
Fetid marigold
Native herb; occasional throughout Illinois on waste ground and in fields.

SYNONYMS: *Boebera papposa* (Vent.)Rydb.
MEDICINE: Used by the Indians of the Missouri River region to treat headache.
REFERENCES: Gilmore 1919:80.

Echinacea pallida (Nutt.)Nutt.
Pale coneflower
Native herb; occasional throughout Illinois in open woods and prairies.

MEDICINE: Possibly used by the Meskwaki for stomach cramps since *E. angustifolia* identified by Smith (1928:212) did not occur in Meskwaki territory. Recent--considered useful for inducing salivation and as an alterative and diaphoretic.
REFERENCES: H. Smith 1928:212; Vogel 1970:357.

Erigeron annuus (L.)Pers.
Daisy fleabane
Native herb; common throughout Illinois in open woods, fields, and waste ground.

MEDICINE: Recent--leaves and flowers considered useful as a stimulant, diuretic, tonic, and astringent.
REFERENCES: Tehon 1951:51; Vogel 1970:305.

Erigeron canadensis L.
Horseweed
Native herb; common throughout Illinois on waste ground.

MEDICINE: Used by the Ojibwa for "female trouble" and stomach pains and by the Meskwaki in steam baths; root used by the Houmas for leukorrhea. Recent--considered useful as a tonic, diuretic, and astringent; leaves and tops used for the treatment of venereal disease.
REFERENCES: Densmore 1928:289; Lewis and Elvin-Lewis 1977:218,326,333; H. Smith 1928:213,272, 1932:429; Tehon 1951:52; Vogel 1970:305.

Erigeron philadelphicus L.
Marsh fleabane
Native herb; common throughout Illinois in open woods, fields, and waste ground.

MEDICINE: Used by the Ojibwa and Meskwaki for colds and fever. Recent--considered useful as a tonic, stimulant, diuretic, and uterine contractant.
SMOKING: Leaves mixed with tobacco or kinnikinnik by the Ojibwa and smoked to attract deer.
REFERENCES: H. Smith 1928:213, 1932:364,429; Vogel 1970:305.

Erigeron strigosus Muhl.
Daisy fleabane
Native herb; common throughout Illinois in open woods, fields, and waste ground.

SYNONYMS: *E. ramosus* (Walt.)B.S.P.
MEDICINE: Used by the Ojibwa to treat headaches.
SMOKING: Flowers smoked as a hunting charm by the Ojibwa.
REFERENCES: H. Smith 1932:364,429; Vogel 1970:305.

Eupatorium maculatum L.
Spotted joe-pye weed
Native herb; common in the northern one-half of Illinois on moist ground.

MEDICINE: Used by the Ojibwa for "medicine" and by the Potawatomi for burns. The plant may cause contact dermatitis in some persons.
REFERENCES: Densmore 1928:289; Lewis and Elvin-Lewis 1977:84; H. Smith 1933:52.

Eupatorium perfoliatum L.
Common boneset
Native herb; common throughout Illinois on wet ground.

MEDICINE: Used by the Menominee for fevers, the Meskwaki for snakebite, and the Ojibwa as a poultice. Recent--leaves and flowers considered useful as a tonic, diaphoretic, emetic, and cathartic. This plant may cause contact dermatitis in some persons.
TECHNOLOGY: Used to make a deer call by the Ojibwa.
REFERENCES: Densmore 1928:289; Gilmore 1933:142; Lewis and Elvin-Lewis 1977:84,374; H. Smith 1923:30, 1928:214.

Eupatorium purpureum L.
Purple joe-pye-weed
Native herb; common throughout Illinois in open woods.

MEDICINE: Used by the Menominee for genitourinary problems and by the Ojibwa for "female troubles."
CEREMONY: Used by the Meskwaki as a love charm and by the Potawatomi for good luck.
REFERENCES: Gilmore 1933:142; Lewis and Elvin-Lewis 1977:218,328; H. Smith 1923:30, 1932:364, 1933:117; Tehon 1951:54; Vogel 1970:328-9.

Eupatorium rugosum Houtt.
White snakeroot
Native herb; common throughout Illinois in woods.

SYNONYMS: *E. urticaefolium* Reich.
MEDICINE: Used by the Meskwaki as a reviver for unconscious persons.
REFERENCES: H. Smith 1928:214.

Eupatorium serotinum Michx.
Late boneset
Native herb; common throughout Illinois in moist open woods and clearings.

MEDICINE: Used by the Houmas to treat typhoid fever.
REFERENCES: Vogel 1970:329.

Gnaphalium obtusifolium L.
Sweet everlasting
Native herb; common throughout Illinois in meadows and open woods.

SYNONYMS: *G. polycephalum* Michx.
MEDICINE: Used by the Meskwaki and Menominee as a reviver for unconscious persons.
SMOKING: Used as a tobacco substitute by the Choctaw. Flowers smoked or smudged by the Potawatomi to drive off evil spirits.
REFERENCES: Epstein 1981:63; H. Smith 1923:30, 1928:214, 1933:48,117.

Grindelia squarrosa (Pursh)Dunal
Gumweed
Introduced herb; occasional in fields and waste ground of the northern one-half of Illinois; rare elsewhere.

MEDICINE: Used by the Indians of the Missouri River Region to treat colic.
REFERENCES: Gilmore 1919:81; Vogel 1970:313.

Helenium autumnale L.
Autumn sneezeweed
Native herb; common throughout Illinois on wet ground.

MEDICINE: Used by the Menominee and Meskwaki to treat head colds. This species is poisonous to livestock and may cause contact dermatitis in some persons.
REFERENCES: Lewis and Elvin-Lewis 1977:57,84; H. Smith 1923:31, 1928:215.

Helianthus annuus L.
Common sunflower
Introduced herb; often cultivated and frequent ruderal throughout Illinois.

FOOD: Plant cultivated for edible seeds by historic and prehistoric Indians as well as being extensively cultivated at the present.
MEDICINE: Used by the Indians of the Missouri River region to treat pulmonary troubles. Recent--considered useful as a diuretic and expectorant.
DYE: Recent--ray flowers used to produce yellow
REFERENCES: Gilmore 1919:77; Lewis and Elvin-Lewis 1977:218; Lust 1974:562; Tehon 1951:63.

Helianthus decapetalus L.
Pale sunflower
Native herb; occasional throughout Illinois in dry open woods.

MEDICINE: Used by the Meskwaki to make poultices for sores.
REFERENCES: H. Smith 1928:215

Helianthus grosseserratus Martens
Sawtooth sunflower
Native herb; common throughout Illinois in prairies, open woods and fence-rows.

MEDICINE: Used by the Meskwaki to make poultices for burns.
REFERENCES: H. Smith 1928:215.

Helianthus strumosus L.
Pale-leaved sunflower
Native herb; common throughout Illinois in open woods.

MEDICINE: The root used for lung trouble by the Meskwaki.
REFERENCES: H. Smith 1928:215.

Helianthus tuberosus L.
Jerusalem artichoke
Native herb; common throughout Illinois in open woods and prairies.

SYNONYMS: *H. tomentosus* Michx., *H. subcanescens* (Gray)E.E. Wats.
FOOD: Tubers eaten raw or cooked by Indians throughout the range of the plant; also cultivated by Historic Indians.
REFERENCES: Densmore 1928:319; Lewis and Elvin-Lewis 1977:218; H. Smith 1928:256, 1933:98; Waugh 1916:120; Yanovsky 1936:61.

Heliopsis helianthoides (L.)Sweet
False sunflower

SYNONYMS: *H. scabra* Dunal
MEDICINE: Used by the Meskwaki to treat lung troubles and by the Ojibwa as a tonic.
REFERENCES: Densmore 1928:289; H. Smith 1928:215.

Hieracium canadense Michx.
Canada hawkweed
Native herb; occasional in dry woods and thickets in the northern quarter of Illinois.

CEREMONY: Flowers used by the Ojibwa as a hunting lure.
REFERENCES: H. Smith 1932:429.

Inula helenium L.
Elecampane
Introduced herb; occurs in fields, waste ground, and open woods; uncommon.

MEDICINE: Recent--the plant has been considered useful as a tonic and stimulant.
DYE: Recent--roots used to produce blue.
REFERENCES: Lust 1974:555; Tehon 1951:67.

Iva annua L.
Sumpweed, marsh-elder
Native herb; occasional throughout Illinois on moist ground.

SYNONYMS: *I. ciliata* Willd.
FOOD: Variety *macrocarpa* domesticated by the Indians of eastern North America for the edible seeds and evidently cultivated into early Historic times. The plant may cause contact dermatitis in some persons.
REFERENCES: Asch and Asch 1977; Black 1963; R. Jackson 1960.

Krigia biflora (Michx.)Nutt.
False dandelion
Native herb; common throughout Illinois on wooded slopes and ridges.

TECHNOLOGY: Used by the Menominee in making deer calls and lures.
REFERENCES: H. Smith 1923:80.

Lactuca canadensis L.
Wild lettuce
Native herb; common throughout Illinois on waste ground.

MEDICINE: Sap used by the Menominee to treat poison ivy and by the Ojibwa for warts.
REFERENCES: Densmore 1928:290; H. Smith 1923:31; Vogel 1970:392.

Lactuca serriola L.
Wild lettuce
Introduced herb; common throughout Illinois on waste ground.

SYNONYMS: *L. scariola* L.
MEDICINE: Used by the Meskwaki to aid lactation in nursing mothers. Recent--considered useful as a sedative.
REFERENCES: H. Smith 1928:215; Tehon 1951:71.

Liatris punctata Hook.
Blazing star
Introduced herb; has been found along railroad tracks in DuPage County; rare.

MEDICINE: Root used by the Meskwaki to treat bladder trouble in women.
REFERENCES: H. Smith 1928:216.

Liatris pycnostachya Michx.
Prairie blazing star
Native herb; common throughout Illinois in prairies.

MEDICINE: Recent--considered useful as a diuretic, diaphoretic, and uterine tonic. Root chewed by early Euro-American inhabitants of New England for venereal disease.
REFERENCES: Lewis and Elvin-Lewis 1977:333; Tehon 1951:72; Vogel 1970:330-32.

Parthenium integrifolium L.
Wild quinine
Native herb; common throughout Illinois in prairies and dry woods.

MEDICINE: Fresh leaves used as a poultice for burns by the Catawbas.
REFERENCES: Lewis and Elvin-Lewis 1977:344; Tehon 1951:85; Vogel 1970:214.

Prenanthes alba L.
White lettuce
Native herb; occasional in woods in the northern one-half of Illinois.

MEDICINE: Herb used by the Ojibwa to treat "female troubles."
REFERENCES: Densmore 1928:291; H. Smith 1932:365; Vogel 1970:222.

Ratibida pinnata (Vent.)Barnh.
Drooping coneflower
Native herb; common in prairies in the northern three-quarters of Illinois.

SYNONYMS: *Lepachys pinnata* (Vent.)T.&G.
MEDICINE: Root used by the Meskwaki to treat toothache.
REFERENCES: H. Smith 1928:216.

Rudbeckia fulgida Ait.
Orange coneflower
Native herb; occurs in woods in Pope County; rare.

MEDICINE: Used by the Cherokee for "medicine."
REFERENCES: Vogel 1970:102.

Rudbeckia hirta L.
Black-eyed susan
Native herb; common throughout Illinois in open woods and fields.

MEDICINE: Root used by the Potawatomi in making a medicinal beverage to treat colds, by the Menominee as a diuretic, and by the Ojibwa as a poultice. The plant has been associated with livestock poisoning.
DYE: Disk florets used by the Potawatomi for yellow.
REFERENCES: Gilmore 1933:143; Lewis and Elvin-Lewis 1977:58; H. Smith 1923:31, 1933:53,117.

Rudbeckia laciniata L.
Goldenglow
Native herb; occasional throughout Illinois on alluvial soil.

FOOD: The young shoots were eaten by the Indians of New Mexico.
MEDICINE: Used by the Ojibwa for burns and indigestion. The plant has been implicated in livestock poisoning.
REFERENCES: Densmore 1928:292; Lewis and Elvin-Lewis 1977:58; Yanovsky 1936:63.

Senecio aureus L.
Golden ragwort
Native herb; occasional throughout Illinois on wet ground.

MEDICINE: Recent--this plant has been considered useful as a diaphoretic, diuretic, tonic, emmenagogue, and uterine contractant. The genus has been associated with livestock poisoning, contains teratogens and carcinogens, and is a possible source of drugs valuable in cancer chemotherapy.
REFERENCES: Lewis and Elvin-Lewis 1977:58,93,120,135,321; Tehon 1951:105.

Senecio vulgaris L.
Common groundsel
Introduced herb; occasional on waste ground in the northern one-half of Illinois.

MEDICINE: Same as *S. aureus* above.
REFERENCES: Tehon 1951:105.

Silphium integrifolium L.
Rosinweed
Native herb; common throughout Illinois in prairie.

MEDICINE: Leaves used by the Meskwaki to make a medicinal beverage for treating bladder trouble. Gummy sap chewed by Indians and early settlers to clean teeth.
REFERENCES: Lewis and Elvin-Lewis 1977:247,267; H. Smith 1928:216; Vogel 1970:248.

Silphium laciniatum L.
Compass-plant
Native herb; common throughout Illinois in prairies.

MEDICINE: Smaller roots boiled and cooled liquid drunk by Meskwaki as an emetic. Recent--considered useful as a treatment for rheumatism, scrofula, and glandular enlargements.
REFERENCES: H. Smith 1928:216; Tehon 1951:106; Vogel 1970:10,248.

Silphium perfoliatum L.
Cup-plant
Native herb; common throughout Illinois in open woods and low ground.

MEDICINE: Root used by the Ojibwa for stomach trouble and hemorrhage, for rheumatism and as an emetic; by the Meskwaki to prevent vomiting during pregnancy. Recent--the gum was used as a styptic and the root as a stimulant, diuretic, and diaphoretic.
REFERENCES: Densmore 1928:292; Gilmore 1919:80; H. Smith 1928:217, 1932:365; Tehon 1951:106; Vogel 1970:104,248.

Solidago flexicaulis L.
Broadleaf goldenrod
Native herb; occasional throughout Illinois in woods.

SYNONYMS: *S. latifolia* L.
MEDICINE: Herb used by the Potawatomi and Ojibwa to treat certain kinds of fevers. Members of the genus *Solidago* appear to be toxic to sheep.
REFERENCES: Densmore 1928:293; Lewis and Elvin-Lewis 1977:58; H. Smith 1933:53.

Solidago gigantea Ait.
Late goldenrod
Native herb; common throughout Illinois on moist ground.

SYNONYMS: *S. serotina* Ait.
MEDICINE: Infusion of flowers used by the Potawatomi and Ojibwa for fevers.
REFERENCES: Densmore 1928:293; H. Smith 1923:31, 1933:53.

Solidago graminifolia (L.)Salisb.
Grass-leaved goldenrod
Native herb; common throughout Illinois on moist ground.

SMOKING: Flowers added to Ojibwa hunting medicine smoked to attract deer.
REFERENCES: Epstein 1981:68.

Solidago juncea Ait.
Early goldenrod
Native herb; common throughout Illinois in open woods, fields, and waste ground.

MEDICINE: Used by the Potawatomi and Ojibwa for "medicine."
REFERENCES: Densmore 1928:293.

Solidago nemoralis Ait.
Field Goldenrod
Native herb; common throughout Illinois in open woods and fields.

MEDICINE: The roots were used by the Houmas for "yellow jaundice."
REFERENCES: Vogel 1970:312-13.

Solidago rigida L.
Rigid goldenrod
Native herb; common throughout Illinois in prairies.

MEDICINE: Infusion of flowers used by the Meskwaki to bathe bee stings; used by the Ojibwa for "medicine."
REFERENCES: Densmore 1928:293; H. Smith 1928:217; Vogel 1970:217.

Solidago speciosa Nutt.
Showy Goldenrod
Native herb; common throughout Illinois in open woods and prairie.

MEDICINE: Root used by the Meskwaki to treat burns and by the Ojibwa for "medicine."
REFERENCES: Densmore 1928:293; H. Smith 1928:218.

Solidago spp.
Goldenrod
Native herb; common in woods and prairies throughout Illinois.

DYE: Recent--flowers used to produce various yellows.
REFERENCES: Brooklyn Botanic Garden 1964:20; Lust 1974:555.

Solidago uliginosa Nutt.
Swamp goldenrod
Native herb; occasional in bogs and swamps in the northern one-half of Illinois.

MEDICINE: Root used by the Ojibwa and Potawatomi as a poultice for boils.
REFERENCES: Densmore 192:293; H. Smith 1933:54.

Solidago ulmifolia Muhl.
Elm-leaved goldenrod
Native herb; common in woods throughout Illinois.

MEDICINE: Herb smudged by the Meskwaki and Ojibwa as a "reviver."
REFERENCES: Densmore 1928:293; H. Smith 1928:218; Vogel 1970:187.

Sonchus arvensis L.
Field sow thistle
Introduced herb; occasional in fields and waste ground in the northern one-half of Illinois.

MEDICINE: Fresh leaves used by the Potawatomi to make a medicinal beverage to treat caked breasts.
CEREMONY: Herb used by the Potawatomi as a deer lure.
REFERENCES: H. Smith 1933:54,117.

Tanacetum vulgare L.
Tansy
Introduced herb; occasional in fields and waste ground throughout Illinois.

SYNONYMS: The native *T. huronense* Nutt. may also have been used.
MEDICINE: Possibly used by the Ojibwa to treat fevers. Recent--leaves and flowers considered useful as a stimulant, abortifacient, and anthelminthic. Tansy oil is toxic in overdose and can cause contact dermatitis in some persons.
TECHNOLOGY: Used as a deer lure by the Ojibwa.
DYE: Recent--leaves used for yellow-green.
REFERENCES: Brooklyn Botanic Garden 1964:67; Densmore 1928:293; Lewis and Elvin-Lewis 1977:58,84; Lust 1974:562; H. Smith 1932:366,429; Tehon 1951:110; Vogel 1970:520.

Taraxacum officinale Weber
Common dandelion
Introduced herb; common throughout Illinois in lawns, fields, and waste ground.

MEDICINE: Root used by the Meskwaki for chest pain, by the Potawatomi as a tonic, and by the Ojibwa for "female trouble." Recent--considered useful as a tonic, diuretic, aperient, and the juice used to remove warts.
DYE: Recent--whole plant used for magenta.
SMOKING: Flowers added to hunting mixture by the Ojibwa.
REFERENCES: Brooklyn Botanic Garden 1964:64; Densmore 1928:293; Epstein 1981:129; Lewis and Elvin-Lewis 1977:218,353; H. Smith 1928:218, 1932:429, 1933:54; Tehon 1951:111.

Xanthium commune Britt.
Cocklebur
Native herb; common throughout Illinois on disturbed ground.

SYNONYMS: *X. italicum* Moretti
FOOD: Ground seeds supposedly eaten in New Mexico. Seeds and seedlings are poisonous to livestock. The plant may cause contact dermatitis in some persons.
REFERENCES: Lewis and Elvin-Lewis 1977:58,74; Yanovsky 1936:63.

BALSAMINACEAE
Balsam Family

Impatiens biflora Walt.
Spotted touch-me-not
Native herb; common on wet soil throughout Illinois.

SYNONYMS: *I. capensis* Meerb.
MEDICINE: Used by the Meskwaki for poultices, by the Potawatomi to treat poison ivy, and by the Ojibwa for headaches and skin rashes.
DYE: Used by the Potawatomi and Ojibwa to obtain a deep yellow or orange. Recent--used for yellow.
REFERENCES: Gilmore 1933:36; Lust 1974:557; H. Smith 1928:205, 1932:425,358, 1933:42,116; Vogel 1970:219.

Impatiens pallida Nutt.
Pale touch-me-not
Native herb; common throughout Illinois on wet soil and in open woods.

MEDICINE: Used by the Ojibwa and Potawatomi for "medicine" and by the Missouri River Indians to treat eczema.
DYE: Used by the Ojibwa, Menominee, and Potawatomi.
REFERENCES: Gilmore 1919:49; Vogel 1970:219.

BERBERIDACEAE
Barberry Family

Berberis canadensis Mill.
American barberry
Native shrub; rare in dry woods and on sandstone cliffs.

MEDICINE: Used by early Euro-Americans to treat fevers and to increase appetite.
REFERENCES: Vogel 1970:344.

Caulophyllum thalictroides (L.)Michx.
Blue cohosh
Native herb; occasional throughout Illinois in rich woods.

MEDICINE: Used by several tribes to treat menstrual disorders; by the Ojibwa as an emetic, for lung trouble, and for cramps; and by the Missouri River Indians for fever. Recent--roots used as an alterative, diuretic, and emmenagogue.
REFERENCES: Densmore 1928:288; Gilmore 1919:31; H. Smith 1923:25, 1928:205, 1932:358, 1933:43; Tehon 1951:34.

Jeffersonia diphylla (L.)Pers.
Twinleaf
Native herb; occasional throughout Illinois in rich woods.

MEDICINE: Used by the Meskwaki to treat eczema. Recent--considered useful as an alterative, antispasmodic, diuretic, diaphoretic, and expectorant; root used as stimulant and tonic.
REFERENCES: Lewis and Elvin-Lewis 1977:374; H. Smith 1928:194; Tehon 1951:69.

Podophyllum peltatum L.
Mayapple
Native herb; common throughout Illinois in open woods.

FOOD: Fruit eaten either raw or cooked by the Iroquois, Meskwaki, and probably others.
MEDICINE: Used by the Meskwaki to treat rheumatism and as a physic. Recent--considered useful as a drastic purgative and hydragogue; plant and unripe fruit are poisonous, teratogenic, and of possible use in cancer chemotherapy.
REFERENCES: Lewis and Elvin-Lewis 1977:31,94,135,281; H. Smith 1928:206; Tehon 1951:88; Waugh 1916:129; Vogel 1970:335-36; Yanovsky 1936:26.

BETULACEAE
Birch Family

Alnus glutinosa (L.)Gaertn.
Black alder
Introduced tree; occasional in northern Illinois along rivers.

MEDICINE: Root included in dentrifice used by early Euro-American settlers.
DYE: Recent--bark used to obtain gray-brown to black and brownish-yellow and leaves for greenish yellow.
REFERENCES: Brooklyn Botanic Garden 1964:68; Lewis and Elvin-Lewis 1977:231,245; Lust 1974:551.

Alnus rugosa (Duroi)Spreng.
Speckled alder
Native shrub; rare in moist thickets in the northeastern portion of Illinois.

SYNONYMS: *A. incana* Tuckerman, *A. incana* (L.)Moench used by H. Smith (1928:206, 1933:43,116).
MEDICINE: Bark and roots used by Ojibwa, Potawatomi and Menominee for "medicine"; by Potawatomi for itching and leukorrhea; and by Meskwaki for sore gums, pyorrhea, and stomach trouble. Bark decoction used by "Indians" for burns and scalds.
DYE: Inner bark used by the Ojibwa to obtain light yellow and by Potawatomi for red and brown.
REFERENCES: Lewis and Elvin-Lewis 1977:273,326,344; H. Smith 1923:26, 1928:197, 1932:425; Vogel 1970:270.

Alnus serrulata (Ait.)Willd.
Smooth alder
Native tree; occasional in the southern one-third of Illinois along rocky streams.

MEDICINE: Recent--bark of stem used as an alterative and astringent.
REFERENCES: Tehon 1951:17.

Betula lutea Michx.
Yellow birch
Native tree in boggy woods in the northeastern portion of Illinois; rare.

FOOD: Sweet sap added to maple sap by Ojibwa and Potawatomi as well as used as a beverage.
MEDICINE: Sap used by Potawatomi to season other medicine; twig used in the Southeast as a chewing stick; boiled bark of *Betula* sp. used by "Indians" to relieve pain of burns, scalds, and wounds.
TECHNOLOGY: Used by Potawatomi for wigwam poles.
REFERENCES: Lewis and Elvin-Lewis 1977:231,344; H. Smith 1932:397, 1933:44; Vogel 1970: 280-82.

Betula nigra L.
River birch
Native tree; local to common in Illinois along rivers and streams.

MEDICINE: Used by the Ojibwa to treat stomach pains and in a salve for ringworm and sores by the Catawbas.
REFERENCES: Densmore 1928:287; Vogel 1970: 280-82.

Betula papyrifera Marsh.
Paper birch
Native tree; rare in northern Illinois in wooded ravines and on low dunes and ridges.

MEDICINE: Sap used for stomach pain by the Ojibwa and to season other medicines by them and the Potawatomi.
TECHNOLOGY: Widespread Indian use for canoes, buckets, kitchen utensils, house covers, and other tools.
DYE: Outer and inner bark used by Ojibwa for light brown and green.
REFERENCES: Densmore 1928:288; Gilmore 1919:23; H. Smith 1928:267, 1932:358,413,425, 1933:43; Vogel 1970:280-82.

Betula pumila L.
Dwarf birch
Native tree; occurs in bogs in northeastern Illinois, rare.

MEDICINE: Cones heated over coals and the smoke inhaled by the Ojibwa for catarrh; tea made for women in childbirth and during menses.
TECHNOLOGY: Stems used for basket ribs and in weaving by the Ojibwa.
REFERENCES: Lewis and Elvin-Lewis 1977:307,323; H. Smith 1932:359,417; Vogel 1970:280-82.

Carpinus caroliniana Walt.
Blue beech
Native tree; common throughout Illinois in moist woods.

TECHNOLOGY: Wood used for wigwam poles by the Ojibwa.
REFERENCES: Gilmore 1933:127.

Corylus americana Walt.
Hazelnut
Native shrub; common throughout Illinois in thickets and dry, disturbed woods.

FOOD: Nuts used by Indians throughtout its range either fresh, cooked with soup, or stored for winter use.
MEDICINE: Used by the Ojibwa as a poultice and by the Menominee to treat for worms.
TECHNOLOGY: Stems used by the Ojibwa for basket ribs and brushes.
DYE: Used by the Ojibwa for black.
REFERENCES: Densmore 1928:289; Gilmore 1919:74; Parker 1910:99; H. Smith 1923:26,63, 1928:256, 1932:359,397,417,425; Yanovsky 1936:17.

Ostrya virginiana (Mill.)K. Koch
Hop hornbeam
Native tree; occasional in woods throughout Illinois.

MEDICINE: Used by the Potawatomi and Ojibwa to treat diarrhea and by the Ojibwa for kidney trouble. Recent--bark and inner bark considered useful as a tonic, alterative, and antiperiodic.
TECHNOLOGY: Wood used for wigwam poles by the Ojibwa.
REFERENCES: Densmore 1928:291, 1929:23; H. Smith 1933:44.

BORAGINACEAE
Borage Family

Cynoglossum officinale L.
Common hound's-tongue
Introduced herb; occasional to common in the northern two-thirds of Illinois in pastures, fields, and along roadsides.

MEDICINE: Recent--leaf and root used as demulcents and sedatives; poisonous to livestock; contains possible carcinogens.
REFERENCES: Lewis and Elvin-Lewis 1977:56,120; Tehon 1951:45.

Hackelia virginiana (L.)I.M. Johnson
Beggar's lice
Native herb; found occasionally in moist or dry woods throughout Illinois.

SYNONYMS: *C. morrisoni* A.DC.
MEDICINE: Used by the Cherokee for "medicine."
REFERENCES: Tehon 1951:102.

Lithospermum canescens Michx.
Hoary puccoon
Native herb; occasional to common throughout Illinois on sandy soil.

CEREMONY: Possibly used ceremonially by the Menominee.
REFERENCES: H. Smith 1923:90.

Lithospermum caroliniense (Walt.)Macm.
Hairy puccoon
Native herb; occasional to common on sandy soil in the northern half of Illinois, less common elsewhere.

DYE: Root used for red by the Ojibwa.
REFERENCES: Densmore 1928:290.

Lithospermum incisum Lehm.
Yellow puccoon
Native herb; occasionally found in the northern half of Illinois; rare elsewhere.

FOOD: Root eaten by Indians in British Columbia.
REFERENCES: Yanovsky 1936:54.

Mertensia virginica (L.)Pers.
Bluebells
Native herb; occasional throughout Illinois in rich woods.

MEDICINE: Recent--herb used as a tonic.
REFERENCES: Tehon 1951:81.

Onosmodium hispidissimum Mack.
Marbleseed
Native herb; uncommon in the western half of Illinois in rocky prairies and woods.

CEREMONY: Used as a love charm by the Ojibwa.
REFERENCES: Densmore 1928:290.

BRASSICACEAE
Mustard Family

Brassica hirta Moench
White mustard
Introduced herb; cultivated and occasionally escaped.

SYNONYMS: *B. alba* (L.)Rabenh.
MEDICINE: Recent--the pungent oil has been used as an emetic, rubefacient, and digestive stimulant. Plants of this genus may be toxic to livestock, cause photodermatitis, or be mutagenic.
REFERENCES: Lewis and Elvin-Lewis 1977:36,81,94,273,278; Tehon 1951:30.

Brassica nigra (L.)Koch
Black mustard
Introduced herb; common throughout Illinois on waste ground.

MEDICINE: Ground seeds used by Meskwaki as a snuff to cure head colds. Recent--seeds chewed for toothache (see also *B. hirta* above).
REFERENCES: Lewis and Elvin-Lewis 1977:94,252,348; H. Smith 1928:219; Tehon 1951:31; Vogel 1970:341-42.

Capsella bursa-pastoris (L.)Medic.
Shepherd's purse
Introduced herb; common throughout Illinois on waste ground.

SYNONYMS: *Bursa bursa-pastoris* (L.)Britt.
MEDICINE: Used by the Meskwaki for "medicine" and by the Ojibwa for dysentery. Recent--considered useful as a tonic, antiscorbutic, and astringent.
REFERENCES: Densmore 1928:288; H. Smith 1923:33, 1928:219; Tehon 1951:32.

Dentaria laciniata Muhl.
Toothwort
Native herb; common throughout Illinois in woods.

FOOD: Roots eaten by the Iroquois.
REFERENCES: Waugh 1916:120; Yanovsky 1936:27.

Erysimum cheiranthoides L.
Wormseed mustard
Introduced herb; occasional in fields and waste ground in the northern half of Illinois.

MEDICINE: Root used by the Ojibwa to treat skin eruptions.
REFERENCES: Densmore 1928:289,350.

Iodanthus pinnatifidus (Michx.)Steud.
Purple rocket
Native herb; occasional throughout Illinois in low woods.

MEDICINE: Herbal poultice applied to the head by the Meskwaki to warm the whole body.
REFERENCES: H. Smith 1928:219.

Lepidium virginicum L.
Peppergrass
Native herb; common throughout Illinois on waste ground.

MEDICINE: Bruised plant and infusion used by the Menominee to treat poison ivy.
REFERENCES: H. Smith 1923:33; Vogel 1970:218-19.

Nasturtium officinale R.Br.
Watercress
Introduced herb; common throughout Illinois in cool springs and branches.

FOOD: Pungent herb edible and often cultivated.
MEDICINE: Herb contains an oil that has been considered useful as an antiscorbutic and blood purifier.
REFERENCES: Tehon 1951:95.

Rorippa islandica (Oeder)Borbas
Marsh yellow cress
Native herb; common throughout Illinois on wet ground and in shallow water.

SYNONYMS: *Radicula palustris* (L.)Moench
FOOD: Plant eaten in the west.
REFERENCES: Yanovsky 1936:27.

CABOMBACEAE
Watershield Family

Brasenia schreberi Gmel.
Watershield
Native herb: in ponds and slow streams; rare throughout Illinois.

MEDICINE: Used by the Meskwaki for stomach ailments and during menses.
REFERENCES: H. Smith 1928:197.

CACTACEAE
Cactus Family

Opuntia compressa (Salisb.)Macbr.
Prickly pear
Native herb; on sandy soil; occasional throughout Illinois.

FOOD: Many similar species were used for food by Historic Indian groups.
REFERENCES: Yanovsky 1936:45–46.

CAMPANULACEAE
Bellflower Family

Campanula americana L.
American bellflower
Native herb; common in woods throughout Illinois.

MEDICINE: Used by the Meskwaki for coughs and consumption.
REFERENCES: H. Smith 1928:206.

Campanula rotundifolia L.
Bellflower
Native herb; occasional in the northern half of Illinois in woods, hill prairies, and on sandstone cliffs; also in Jackson County.

SYNONYMS: *C. intercedens* Witasek
MEDICINE: Root used by the Ojibwa to treat lung troubles and ear diseases.
REFERENCES: Densmore 1928:288; Gilmore 1933:142; H. Smith 1932:360.

Lobelia cardinalis L.
Cardinal flower
Native herb; found occasionally on wet ground throughout Illinois.

MEDICINE: Used by the Cherokee to treat worms; plant is poisonous.
CEREMONY: Root used by both the Meskwaki and Indians of the Missouri River region as love charms; strewn on graves and used to ward off storms by the Meskwaki.
SMOKING: Herb smoked ceremonially by the Menominee.
REFERENCES: Epstein 1981:38; Gilmore 1919:77; Lewis and Elvin-Lewis 1977:56, 329; H. Smith 1928:231,273; Vogel 1970:330–32.

Lobelia inflata L
Indian tobacco
Native herb; occasional throughout Illinois in woods, fields, and on disturbed ground.

MEDICINE: Used by "Indians" as purgative, astringent, or emetic. Recent--considered useful as an antispasmodic in laryngitis and spasmodic asthma and as an expectorant or emetic; the plant is poisonous.

SMOKING: Possibly used by some "Indians" as a tobacco substitute or adulterant.
REFERENCES: Epstein 1981:40; Lewis and Elvin-Lewis 1977:56; Tehon 1951:74; Vogel 1970:330-32.

Lobelia spicata Lam.
Spiked lobelia
Native herb; occasional throughout Illinois in dry woods and prairies.

MEDICINE: Plant poisonous; used by Euro-Americans as a diuretic.
REFERENCES: Lewis and Elvin-Lewis 1977:56; Vogel 1970:330-32.

Lobelia syphilitica L.
Blue cardinal-flower
Native herb; occasional to common on wet ground throughout Illinois.

MEDICINE: Used by the Cherokee and Iroquois to treat venereal disease; the plant is poisonous.
CEREMONY: Used by the Meskwaki as a love charm.
REFERENCES: Lewis and Elvin-Lewis 1977:56; H. Smith 1928:231,273; Vogel 1970:330-32.

Specularia perfoliata A.DC.
Venus' looking-glass
Native herb; common throughout Illinois on dry, often disturbed soil.

MEDICINE: Used by the Meskwaki as an emetic.
SMOKING: Smoked ceremonially by the Meskwaki.
REFERENCES: Epstein 1981:42; H. Smith 1928:206,272.

CANNABINACEAE
Hemp Family

Cannabis sativa L.
Marijuana, hemp
Introduced herb; occasional throughout Illinois on waste ground.

MEDICINE: Recent--used as a narcotic, analgesic, and sedative.
TECHNOLOGY: Introduced into North America and once widely grown as a commercial source of hemp fibers.
REFERENCES: Tehon 1951:32.

Humulus lupulus L.
Hops
Native herb; Occurs occasionally throughout Illinois in thickets, dry woods and on slopes.

SYNONYMS: *H. americanum* Nutt.
MEDICINE: Used to make a medicinal beverage for indigestion by the Ojibwa; as a tonic, diuretic, and sedative by the Meskwaki; to treat fevers and intestinal pains by the Indians of the Missouri River region. Recent--flowers used as tonic, sedative, and hypnotic and to flavor beer. The plant can cause contact dermatitis.
REFERENCES: Gilmore 1919:25; Lewis and Elvin-Lewis 1977:82; H. Smith 1928:250; Tehon 1951; Vogel 1970:318.

CAPPARIDACEAE
Caper Family

Polanisia dodecandra (L.)DC. var. *trachysperma* (T.&G.)Iltis
Clammyweed
Native herb; occurs on sandy soil throughout Illinois; uncommon.

SYNONYMS: *P. trachysperma* Torr. & Gray
FOOD: Plants eaten by Indians in the West.
MEDICINE: Recent--the root and plant recognized as active anthelminthics.
REFERENCES: Lewis and Elvin-Lewis 1977:291; Yanovsky 1936:239.

CAPRIFOLIACEAE
Honeysuckle Family

Diervilla lonicera Mill.
Bush honeysuckle
Native shrub; occasionally occurs in sandy woods in the northern quarter of Illinois.

MEDICINE: Used by the Menominee to treat senility and as a diuretic; by the Ojibwa for stomach and urinary troubles and for constipation; and by the Meskwaki to treat gonorrhea.
REFERENCES: Densmore 1928:289; Gilmore 1933:141; H. Smith 1923:27, 1928:206, 1932:360, 1933:45-46,69.

Linnaea americana Forbes
Twinflower
Native herb; occurs in bogs in the northeastern portion of Illinois, rare.

SYNONYMS: *L. borealis* L.
MEDICINE: Used by the Potawatomi for "female troubles."
REFERENCES: H. Smith 1933:45.

Lonicera dioica L.
Red honeysuckle
Native vine; occurs in woods in the northeastern portion of Illinois; uncommon.

MEDICINE: Used by the Meskwaki to treat for worms and by the Ojibwa for lung trouble.
REFERENCES: Densmore 1928:290; H. Smith 1928:207.

Lonicera flava Sims
Yellow honeysuckle
Native shrub; occurs in rocky woods and on cliffs in the southern tip of Illinois, rare.

MEDICINE: Used by the Meskwaki for "internal troubles." The plant referred to by Smith was more likely to have been *L. prolifera* (Kirch.)Rehder since *L. flava* does not occur in Meskwaki territory.
REFERENCES: H. Smith 1928:195.

Sambucus canadensis L.
Elderberry
Native shrub; common throughout Illinois on moist ground, in woods and along streams and lakes.

FOOD: Berries eaten fresh or cooked by many Historic tribes and a beverage made from the blossoms.
MEDICINE: Used to make a medicinal beverage by the Ojibwa, Sauk-Fox and Iroquois and as an emetic by the Ojibwa. Bark tea used by Meskwaki to aid childbirth. Recent--the bark used as a cathartic, flowers mildly astringent. Plant poisonous except for the blossoms and ripe fruit.
REFERENCES: Gilmore 1919:115, 1933:142; Lewis and Elvin-Lewis 1977:50,322; H. Smith 1923:27, 1928:256; Tehon 1951:101; Vogel 1970:301-2; Waugh 1916:128; Yanovsky 1936:58.

Sambucus pubens Michx.
Red-berried elder
Native shrub occurring in moist woods in northeastern Illinois; uncommon.

FOOD: Berries supposedly eaten by the Ojibwa and a beverage made from the roots by Historic tribes outside of the Midwest.
MEDICINE: Bark used for medicine by the Ojibwa, Menominee, and Potawatomi. The plant poisonous except for the flowers and ripe fruit.
REFERENCES: Lewis and Elvin-Lewis 1977:50; Yanovsky 1936:58.

Symphoricarpos albus (L.)Blake
Garden snowberry
Introduced shrub; cultivated and occasionally escaped.

MEDICINE: Root used for medicine by the Ojibwa.
REFERENCES: H. Smith 1932:361.

Symphoricarpos occidentalis Hook.
Wolfberry
Native shrub; occasional in the northern one-fifth of Illinois on dry, open ground.

FOOD: Fruit eaten by the Sioux.
MEDICINE: Used by the Ojibwa to cleanse afterbirth and to enable convalescence.
REFERENCES: H. Smith 1928:207, 1932:361; Yanovsky 1936:58.

Symphoricarpos orbiculatus Moench
Coralberry
Native shrub; occasional throughout Illinois in disturbed woods.

SYNONYMS: *S. symphoricarpos* (L.)Macm.
MEDICINE: Used by the Indians of the Missouri River region to treat eye problems.
REFERENCES: Gilmore 1919:64.

Triosteum perfoliatum L.
Late horse gentian
Native herb; occasional throughout Illinois in dry woods and thickets.

MEDICINE: Used by the Meskwaki to treat rattlesnake bites.
REFERENCES: H. Smith 1928:207; Vogel 1970:307,323.

Viburnum acerifolium L.
Maple-leaved viburnum
Native shrub; occasionally occurs in moist woods in the northeastern one-fourth of Illinois.

MEDICINE: Medicinal beverage used by the Menominee to treat cramps or colic and by the Ojibwa as an emetic. Recent--bark of root found to contain resins considered useful as antispasmodic, nervine, tonic, astringent, and uterine sedative.
SMOKING: Included in smoking mixture by the Omaha and Delaware.
REFERENCES: Densmore 1928:294; Epstein 1981:45; H. Smith 1923:29; Tehon 1951:119.

Viburnum lentago L.
Nannyberry
Native shrub; occasional in moist woods in the northern half of Illinois.

FOOD: Fruit eaten by Iroquois and Missouri River Indians.
MEDICINE: Bark used by the Ojibwa as a diuretic and for dysuria.
REFERENCES: Gilmore 1919:142; H. Smith 1932:361.

CARYOPHYLLACEAE
Pink Family

Lychnis alba Mill.
White campion
Introduced herb; common in the nothern two-thirds of Illinois on waste ground.

MEDICINE: Used as a physic by the Ojibwa.
REFERENCES: H. Smith 1932:361.

Saponaria officinalis L.
Bouncing bet, soapwort
Introduced herb; common on waste ground throughout Illinois.

MEDICINE: Rootstock used by early Euro-Americans as an alterative, poultice to remove discoloration of black or bruised eye, or as a soap substitute.
REFERENCES: Lewis and Elvin-Lewis 1977:224; Tehon 1951:102.

Silene noctiflora L.
Night-flowering catchfly
Introduced herb; occasional in the southern half of Illinois on waste ground.

MEDICINE: Used for "medicine" by the Menominee.
REFERENCES: H. Smith 1923:28.

Silene stellata (L.)Ait.f.
Starry campion
Native herb; occurs occasionally in woods throughout Illinois.

MEDICINE: Used for infections by the Meskwaki.
REFERENCES: H. Smith 1928:208.

Stellaria media (L.)Cyrill.
Common chickweed
Introduced herb; common on waste ground throughout Illinois.

MEDICINE: Herb used to treat sore eyes by the Ojibwa.
REFERENCES: Densmore 1928:293; Tehon 1951:109.

CELASTRACEAE
Bittersweet Family

Celastrus scandens L.
Bittersweet
Native vine; occasional in woods throughout Illinois.

FOOD: Inner bark and twigs supposedly eaten by the Indians of Minnesota and Wisconsin after lengthy boiling. The leaves, bark, and fruit are poisonous.
MEDICINE: Used by the Ojibwa for stomach trouble and skin eruptions, as a physic, and to treat leukorrhea. Recent--bark of root and stem considered useful as a diaphoretic, alterative, diuretic, narcotic, and insecticide.
REFERENCES: Densmore 1928:288; Gilmore 1919:102, 1933:135; Lewis and Elvin- Lewis 1977:48,326; H. Smith 1923:63, 1928:208, 1932:362, 1933:97.

Euonymus atropurpureus Jacq.
Wahoo
Native shrub; occasional in woods throughout Illinois.

MEDICINE: Poultice made of inner bark used by the Meskwaki for facial sores and a decoction used as an eye lotion; used by the Indians of the Missouri River region and the Winnebago for uterine problems. Recent--the bark was used as a mild purgative. The bark, leaves, and fruit are probably all poisonous.
SMOKING: Bark used in a smoking mixture by the Senecas and Delaware Indians.
REFERENCES: Epstein 1981:46; Gilmore 1919:50; Tehon 1951:55; Vogel 1970:387-88.

CHENOPODIACEAE
Goosefoot Family

Chenopodium album L.
Lamb's quarters
Introduced herb; common throughout Illinois on waste ground.

FOOD: Greens and seeds edible but there is no evidence they were eaten by American Indians.
MEDICINE: Pollen sometimes used as a hay fever antigen.
DYE: Used by the Indians of the Missouri River region to paint bows and arrows green.
REFERENCES: Gilmore 1919:26; Tehon 1951:38.

Chenopodium ambrosioides L. var. *anthelminticum* (L.)Gray
Mexican tea, wormwood
Introduced herb; occasionally found on waste ground in the southern half of Illinois.

MEDICINE: Recent--flowers and seeds contain a volatile oil formerly used as a vermifuge and to treat amoebic dysentery; the theraputic dose is dangerously close to the minimum toxic level, however.
REFERENCES: Lewis and Elvin-Lewis 1977:291-92; Tehon 1951:38.

Chenopodium bushianum Aellen
Goosefoot
Native herb; occurs occasionally on dry soil throughout Illinois.

FOOD: Seeds possibly used aboriginally for food; the plant may have been cultivated to some extent although there is no historic record of its use.
REFERENCES: Asch and Asch 1977.

Chenopodium capitatum (L.)Aschers.
Strawberry blite
Introduced herb; occurs on sandy soil; rare.

MEDICINE: Used by the Potawatomi for lung congestion.
DYE: Used by the Potawatomi for red paint.
CEREMONY: Used by the Potawatomi for dream dance paint.
REFERENCES: H. Smith 1933:47,117.

Cycloloma atriplicifolium (Spreng.)Coult.
Winged pigweed
Native herb; occasional throughout Illinois on sandy soil.

FOOD: The seeds were eaten by Historic Indians in New Mexico.
REFERENCES: Yanovsky 1936:22.

CISTACEAE
Rockrose Family

Helianthemum canadense (L.)Michx.
Frostweed
Native herb; occasional in sandy woods and prairies in the northern half of Illinois.

MEDICINE: Recent--used as a mild tonic and astringent.
REFERENCES: Tehon 1951:63; Vogel 1970:358.

CONVOLVULACEAE
Morning-glory Family

Cuscuta glomerata Choisy
Dodder
Native herb parasitic on several composites that grow in low, wet areas; occasional throughout Illinois.

DYE: Used by the Indians of the Missouri River region for orange.
REFERENCES: Gilmore 1919:58.

Ipomoea pandurata (L.)Meyer
Wild sweet-potato
Native herb; occasional to common in disturbed areas and thickets in the southern one-half of Illinois.

FOOD: Root eaten in times of famine by southern Indians.
MEDICINE: Root used by Carolina Indians for kidney trouble and rheumatism and as a purgative. Recent--root used as a powerful cathartic and purgative.
REFERENCES: Tehon 1951:68; Vogel 1970:324-25; Yanovsky 1936:53.

CORNACEAE
Dogwood Family

Cornus alternifolia L.
Alternate-leaved dogwood
Native shrub; occurs in moist woods in the northern one-half of Illinois; uncommon.

MEDICINE: Bark used by the Ojibwa and Potawatomi to make an eyewash and by the Ojibwa as an emetic.
TECHNOLOGY: Twigs used by the Ojibwa as a thatch and to attract muskrats.
SMOKING: Bark smoked by the Ojibwa and Menominee.
REFERENCES: Densmore 1928:288; Gilmore 1933:138; H. Smith 1923:32,80, 1932:366,417,429, 1933:54; Vogel 1970:301.

Cornus canadensis L.
Bunchberry
Native herb; occurs in boggy woods in the northeastern portion of Illinois, rare.

FOOD: Fruit eaten fresh by Indians in the northern states and Canada.
MEDICINE: Root used by Ojibwa to treat colic in babies.
REFERENCES: Densmore 1928:321; H. Smith 1932:366, 1933:98; Yanovsky 1936:49.

Cornus foemina Mill.
Stiff dogwood
Native shrub; occasionally occurs in swamps and low woods in the southern one-half of Illinois.

SYNONYMS: *C. stricta* Lam.
MEDICINE: Decoction of root bark used by the Houmas for fever and malaria.
REFERENCES: Vogel 1970:301.

Cornus florida L.
Flowering dogwood
Native tree; occasional to common in rocky woods and low woods and on slopes in the southern one-half of Illinois.

MEDICINE: Used by the Ojibwa to treat coughs and fevers. Recent--flowers, fruit, and bark steeped, making a bitter drink used for fevers and chills; twigs used as chewing sticks; bark considered useful as a tonic and astringent.
REFERENCES: Gilmore 1933:139; Lewis and Elvin-Lewis 1977:168,230; Tehon 1951:44; Vogel 1970:299-301.

Cornus racemosa Lam.
Gray dogwood
Native shrub; occasional to common in moist woods, prairies, and along roadsides throughout Illinois.

SYNONYMS: *C. paniculata* L'her.
MEDICINE: Bark used by the Ojibwa and Meskwaki to treat flux and by the Meskwaki also for toothache, neuralgia, consumption, and as a reviver.
SMOKING: Bark smoked by the Ojibwa and Meskwaki.
REFERENCES: H. Smith 1928:272, 1932:367,418; Vogel 1970:301.

Cornus rugosa Lam.
Round-leaved dogwood
Native shrub; occurs in woods in the northern one-fourth of Illinois; uncommon.

SMOKING: Bark smoked by the Ojibwa.
REFERENCES: Densmore 1928:288

Cornus stolonifera Michx.
Red osier
Native shrub; occasional in open marshes and fens in the northern one-half of Illinois.

FOOD: Fruit eaten by the Indians of the Missouri River Region.
MEDICINE: Root and bark used by the Potawatomi for diarrhea and by the Ojibwa for eye and skin problems; possibly used also by the Menominee.
DYE: Used by the Ojibwa.
SMOKING: Red bark smoked by the Ojibwa, Potawatomi, and Indians of the Missouri River region ceremonially and socially.
REFERENCES: Densmore 1928:288; Gilmore 1919:56, 1933:138; H. Smith 1923:32, 1933:55,117; Vogel 1970:301; Yanovsky 1936:49.

CRASSULACEAE
Stonecrop Family

Penthorum sedoides L.
Ditch stonecrop
Native herb; common throughout Illinois on wet ground.

MEDICINE: Seeds used by the Meskwaki to make a cough syrup.
REFERENCES: H. Smith 1928:219.

CUCURBITACEAE
Gourd Family

Cucurbita foetidissima H.B.K.
Missouri gourd, buffalo gourd
Introduced herb; occasional along railroads.

FOOD: Bitter, young fruit boiled or dried for winter food; seeds roasted or boiled by Indians of the Southwest; root may also have been used for food.
TECHNOLOGY: Fruit pulp used as a detergent for cleaning and scouring by southwestern Indians; dried fruits also used as containers.
REFERENCES: Cutler and Whitaker 1961:473-74; Yanovsky 1936:59.

Echinocystis lobata (Michx.)T.&G.
Wild balsam-apple
Native herb; occasional on moist soil in the northern half of Illinois.

SYNONYMS: *Micrampelis lobata* (Michx.)Greene
MEDICINE: Pounded roots used as a headache poultice by the Meskwaki and Menominee and by the Ojibwa as a tonic and for stomach troubles.
TECHNOLOGY: Seeds used as beads by the Missouri River Indians.
REFERENCES: Gilmore 1919:77; H. Smith 1923:33, 1928:220, 1932:367.

DROSERACEAE
Sundew Family

Drosera rotundifolia L.
Round-leaved sundew
Native herb; occurs in bogs in the northeastern portion of Illinois; rare.

MEDICINE: Recent--the plant was considered useful as a pectoral.
DYE: Recent--plant used for yellow.
REFERENCES: Lust 1974:562; Tehon 1951:50.

EBENACEAE
Ebony Family

Diospyros virginiana L.
Persimmon
Native tree; occasional to common in woods in the southern two-thirds of Illinois.

FOOD: Fruit eaten by Indians throughout the range of the plant species.
MEDICINE: Boiled fruit used by the Cherokee to treat bloody bowel discharges; bark infusion used by the Catawbas for thrush in babies. Recent--tannin in unripe fruit has been considered useful as an astringent and tonic.
REFERENCES: Lewis and Elvin-Lewis 1977:263,285; Tehon 1951:49; Vogel 1970:345-46; Yanovsky 1936:52.

ELAEAGNACEAE
Oleaster Family

Shepherdia canadensis (L.)Nutt.
Buffalo-berry
Native shrub; occurs on dry bluffs and banks or ravines near Lake Michigan; rare.

SYNONYMS: *Lepargyrea canadensis* (L.)Greene
FOOD: Fruit eaten fresh, dried, or made into preserves in the Northwest.
REFERENCES: Yanovsky 1936:46.

ERICACEAE
Heath Family

Andromeda glaucophylla L.
Bog-rosemary
Native herb; occurs in bogs in the northeastern portion of Illinois; rare.

FOOD: Leaves and twigs boiled to make a tealike beverage by the Ojibwa.
REFERENCES: Yanovsky 1936:50.

Arctostaphylos uva-ursi (L.)Spreng.
Bearberry
Native shrub; occurs on sand dunes and in black oak woods in the northern quarter of Illinois; rare.

FOOD: Berries eaten in the Northeast.
MEDICINE: Used as a seasoner for female medicines by the Menominee and for headaches by the Ojibwa. Recent--considered useful as an astringent, tonic, and diuretic.
TECHNOLOGY: Used as a utility plant by the Ojibwa.
DYE: Recent--plant used for green.
CEREMONY: Root smoked by the Ojibwa to attract game.
SMOKING: Leaves used in mixtures smoked by many tribes.
REFERENCES: Brooklyn Botanic Garden 1964:67,70; Densmore 1928:287; Gilmore 1919:56; Lewis and Elvin-Lewis 1977:315; Lust 1974:551; H. Smith 1923:35, 1933:118; Tehon 1951:24; Vogel 1970:200; Yanovsky 1936:50.

Chamaedaphne calyculata (L.)Moench
Leatherleaf
Native shrub; occurs in swamps and bogs in northeastern Illinois; rare.

FOOD: Fresh or dried leaves used by the Ojibwa for making a tealike beverage.
MEDICINE: Leaves used by the Potawatomi for making a medicinal beverage to treat fevers.
REFERENCES: Lewis and Elvin-Lewis 1977:389; H. Smith 1932:400, 1933:56; Yanovsky 1936:50.

Chimaphila umbellata (L.)Bart.
Pipsissewa
Native herb; occurs in dry woods in the northeastern portion of Illinois; rare.

SYNONYMS: *Chimaphila corymbosa* Pursh
MEDICINE: Medicinal beverage used by the Ojibwa for stomach troubles and gonorrhea; by the Menominee as a seasoner for other medicines, and by the Catawbas for backache. Recent--considered useful in treating typhus, as an astringent, tonic, antidiuretic, and for bladder stones; exhibits experimental hypoglycemic activity.
REFERENCES: Gilmore 1933:138; Lewis and Elvin-Lewis 1977:167,218,313,316; H. Smith 1923:35, 1932:368; Vogel 1970:349-50.

Epigaea repens L.
Trailing arbutus
Native herb; occurs in woods in northern Illinois; rare.

MEDICINE: Sometimes used as a substitute for *Arctostaphylos uva-ursi.*
CEREMONY: Tribal flower of the Forest Potawatomi.
REFERENCES: H. Smith 1933:118; Tehon 1951:50.

Gaultheria procumbens L.
Checkerberry, wintergreen
Native herb; occurs on sandy soil of woods and bogs in the northeastern portion of Illinois; rare.

FOOD: Spicy fruit eaten and leaves used for making beverage in the Northeast.
MEDICINE: Leaf tea used by Menominee and Ojibwa for treating rheumatism and colds and by the Potawatomi to break fevers. Recent--considered useful as an aromatic stimulant, flavoring agent, and treatment for rheumatic fever; twigs chewed to prevent dental caries; source of salicylic acid.
REFERENCES: Densmore 1928:317; Lewis and Elvin-Lewis 1977:151, 232, 245, 389; H. Smith 1923:35, 1932:369,400, 1933:56; Tehon 1951:59; Vogel 1970:393-95; Yanovsky 1936:50.

Gaylussacia baccata (Wang.)K.Koch
Black huckleberry
Native shrub; occasional in rocky woods and on hillsides in northern Illinois.

SYNONYMS: *G. resinosa* (Ait.)T.&G.
FOOD: Sweet fruit eaten by the Indians of eastern North America.
REFERENCES: Yanovsky 1936:50.

Monotropa uniflora L.
Indian-pipe
Native herb; occasional in rich woods throughout Illinois.

MEDICINE: Root used by the Potawatomi to treat "female troubles." Recent--the juice was used to treat inflamed eyes during the 19th century.
REFERENCES: Lewis and Elvin-Lewis 1977:224; H. Smith 1933:57.

Pyrola americana Sweet
Shinleaf, wild lily-of-the-valley
Native herb; occurs on shaded, woody slopes in Ogle County; very rare.

SYNONYMS: *P. rotundifolia* L. var. *americana* (Sweet)Fern.
MEDICINE: Used by "Indians" as a sudorific, an astringent, for wounds, and for breast diseases.
CEREMONY: Leaf tea drunk by the Ojibwa as a good luck potion before the start of a hunt.
REFERENCES: Lewis and Elvin-Lewis 1977:342; H. Smith 1932:430; Vogel 1970:394.

Pyrola elliptica Nutt.
Shinleaf
Native herb; occurs in moist woods in the extreme northern portion of Illinois; rare.

MEDICINE: Used by the Ojibwa for making poultices.
REFERENCES: Gilmore 1933:138.

Vaccinium angustifolium Ait.
Low-bush blueberry
Native shrub; occasional in sandy, open woods, dunes and bogs in the northern one-fourth of Illinois.

FOOD: Berries eaten by Indians of the Northeast.
MEDICINE: Beverage made from leaves used by the Ojibwa as a blood purifier and to treat insanity.
REFERENCES: Densmore 1928:231,294; H. Smith 1923:66, 1932:369,401; Yanovsky 1936:51.

Vaccinium arboreum Marsh.
Farkleberry
Native shrub; locally abundant on sandstone cliffs in extreme southern Illinois.

MEDICINE: Recent--fruit and root bark used to treat dysentery.
REFERENCES: Lewis and Elvin-Lewis 1977:285.

Vaccinium corymbosum L.
High-bush blueberry
Native shrub; occurs locally in swamps and bogs in the northeastern portion of Illinois.

FOOD: Berries eaten by Indians of the Northeast.
REFERENCES: Yanovsky 1936:51.

Vaccinium macrocarpon Ait.
American cranberry
Native shrub; occurs in bogs in Cook, Lake, McHenry, and Will counties.

SYNONYMS: *Oxycoccus macrocarpus* (Ait.)Pursh
FOOD: Cooked berries eaten by the Ojibwa and Iroquois.
REFERENCES: Densmore 1928:321; Waugh 1916:128; Yanovsky 1936:51.

Vaccinium myrtilloides Michx.
Canada blueberry
Native shrub; occurs in tamarack bogs and on sandy or rocky slopes in Lake, LaSalle, Ogle, and Winnebago counties.

FOOD: Berries eaten by northern Indians.
MEDICINE: Root used for medicine by the Potawatomi.
REFERENCES: H. Smith 1933:57,99; Yanovsky 1936:51.

Vaccinium vacillans Torr.
Low-bush blueberry
Native shrub; occasional in the southern one-sixth of Illinois; rare elsewhere.

FOOD: Many similar species were used for food by Historic Indian groups.
REFERENCES: Yanovsky 1936:51.

EUPHORBIACEAE
Spurge Family

Chamaesyce maculata (L.)Small
Nodding spurge
Native herb; common throughout Illinois in fields and along roadsides.

SYNONYMS: *Euphorbia maculata* L.
MEDICINE: Used by the Houmas for "medicine."
REFERENCES: Vogel 1970:324.

Chamaesyce serpyllifolia (Pers.)Small
Spurge
Native herb; occurs in sandy areas in the northeastern portion of Illinois; rare.

MEDICINE: Used by the Missouri River Indians to induce milk flow in nursing mothers and to treat dysentery in children.
REFERENCES: Gilmore 1919:47.

Croton texensis (Klotzsch)Muell.-Arg.
Texas croton
Introduced herb; occurs on dry soil in Menard county, rare.

MEDICINE: Used by the Missouri River Indians to treat sick babies. The seeds are poisonous and the plant can cause an allergic reaction.
REFERENCES: Gilmore 1919:47; Lewis and Elvin-Lewis 1977:37,86.

Euphorbia corollata L.
Flowering spurge
Native herb; common throughout Illinois in prairies, woods, fields, and along roadsides.

MEDICINE: Root used as a physic by the Ojibwa and a cathartic by the Meskwaki. Recent--plant considered useful as an emetic, cathartic, and diuretic. Plant and seeds are poisonous and carcinogenic; skin contact causes allergic reactions in some persons.
REFERENCES: Lewis and Elvin-Lewis 1977:37,86; H. Smith 1928:220, 1932:369; Vogel 1970:323-24.

FAGACEAE
Beech Family

Castanea dentata (Marsh.)Borkh.
Chestnut
Native tree; in rocky woods in the southern tip of Illinois; probably extinct.

FOOD: Crushed nuts eaten or boiled for oil by the Iroquois and undoubtedly other groups.
MEDICINE: Medicinal beverage made from dried leaves used to treat whooping cough by the Mohegan. Recent--dried leaves considered useful as a tonic and astringent.
REFERENCES: Vogel 1970:290-91; Yanovsky 1936:17.

Fagus grandifolia Ehrh.
Beech
Native tree; occasional in rich woods in the southern one-fourth of Illinois.

SYNONYMS: *Fagus americana* Sweet, *F. ferruginea* Ait.
FOOD: Nuts eaten fresh or stored for winter use; swelling buds eaten by many groups.
MEDICINE: Leaves used by the Potawatomi to treat burns and by the Ojibwa for pulmonary trouble. Recent--leaves and bark considered useful as a tonic, astringent, and a source of creosote which is an antiseptic and expectorant. The nuts occasionally cause gastrointestinal distress.
TECHNOLOGY: Wood used to make bowls by the Potawatomi.
REFERENCES: Gilmore 1933:128; Lewis and Elvin-Lewis 1977:32, 251, 300, 365; Parker 1910:99; H. Smith 1923:36, 1928:66, 1932:401, 1933:58,100,113; Vogel 1970:219; Yanovsky 1936:18.

Quercus alba L.
White oak
Native tree; common throughout Illinois in upland woods.

FOOD: Acorns eaten in Minnesota and Wisconsin, by the Iroquois, and undoubtedly elsewhere.
MEDICINE: Cambium used as an astringent by the Menominee and made into a medicinal beverage by the Meskwaki to treat lung trouble. Decoction of boiled bark drunk by Iroquois for bleeding hemorrhoids and by Indians and settlers for dysentery. Leaves and acorns of the genus are poisonous to livestock in excessive amounts; the tannins are carcinogenic.
TECHNOLOGY: Wood made into awls and other tools by the Ojibwa and probably other groups.
DYE: Bark used for brown.
REFERENCES: Lewis and Elvin-Lewis 1977:32,245,285,286,293; Lust 1974:564; H. Smith 1923:36,66, 1928:221,257, 1932:402,418; Vogel 1970:342-43; Yanovsky 1936:18.

Quercus bicolor Willd.
Swamp white oak
Native tree; occasional throughout Illinois on alluvial soil.

FOOD: Acorns eaten by the Iroquois and probably others.
REFERENCES: Waugh 1916:123, Yanovsky 1936:18.

Quercus coccinea Muench.
Scarlet oak
Native tree; occasional in upland woods in the southern one-third of Illinois.

FOOD: Many similar species were used for food by Historic Indian groups.
REFERENCES: Yanovsky 1936:18-19.

Quercus ellipsoidalis E.J.Hill
Hill's Oak
Native tree; occasional in upland woods in the northern one-quarter of Illinois.

FOOD: Roasted and ground acorns used by the Menominee to make a beverage.
REFERENCES: H. Smith 1923:66; Yanovsky 1936:18.

Quercus falcata Michx.
Spanish oak
Native tree; occasional in woods in the southern one-fifth of Illinois.

FOOD: Many similar species were used for food by Historic Indian groups.
REFERENCES: Yanovsky 1936:18-19.

Quercus imbricaria Michx.
Shingle oak
Native tree; occasional to common in woods throughout Illinois.

FOOD: Many similar species were used for food by Historic Indian groups.
REFERENCES: Yanovsky 1936:18-19.

Quercus lyrata Walt.
Overcup oak
Native herb; occasional in swamps and bottomlands in the southern one- half of Illinois.

FOOD: Many similar species were used for food by Historic Indian groups.
REFERENCES: Yanovsky 1936:18-19.

Quercus macrocarpa Michx.
Bur oak
Native tree; occasional to common in bottomland woods throughout Illinois.

FOOD: Acorns eaten in Nebraska, North and South Dakota, Minnesota, Wisconsin, and undoubtedly elsewhere.
MEDICINE: Leaves used by the Ojibwa as an astringent and for wounds; wood and inner bark used by the Meskwaki as a vermifuge.
DYE: Bark used for dye by the Ojibwa.

REFERENCES: Densmore 1928:292; Gilmore 1919:75; H. Smith 1928:221, 1932:369,402,425; Vogel 1970:200; Yanovsky 1936:19.

Quercus marilandica Muench.
Blackjack oak
Native tree; common in upland woods and on bluffs in the southern three-fourths of Illinois.

FOOD: Many similar species were used for food by Historic Indian groups.
REFERENCES: Yanovsky 1936:18-19.

Quercus michauxii Nutt.
Basket oak
Native tree; occasional in bottomland woods in the southern one-third of Illinois.

FOOD: Many similar species were used for food by Historic Indian groups.
REFERENCES: Yanovsky 1936:18-19.

Quercus muhlenbergii Engelm.
Yellow chestnut oak
Native tree; occasional throughout Illinois in woods.

FOOD: Many similar species were used for food by Historic Indian groups.
REFERENCES: Yanovsky 1936:18-19.

Quercus palustris Muench.
Pin oak
Native tree; occasional to common throughout Illinois in woods.

FOOD: Many similar species were used for food by Historic Indian groups.
REFERENCES: Yanovsky 1936:18-19.

Quercus phellos L.
Willow oak
Native tree; occasional in moist woods or swamps in the southern tip of Illinois.

FOOD: Many similar species were used for food by Historic Indian groups.
REFERENCES: Yanovsky 1936:18-19.

Quercus prinus L.
Rock chestnut oak
Native tree; occurs in rocky woods in southern Illinois; rare.

SYNONYMS: *Q. montana* Willd.
FOOD: Acorns eaten by the Iroquois.
REFERENCES: Waugh 1916:123; Yanovsky 1936:19.

Quercus rubra L.
Red oak
Native tree; common throughout Illinois in upland woods.

SYNONYMS: *Q. borealis* Michx., *Q. borealis* Michx. var. *maxima* (Marsh.)Ashe
FOOD: Acorns eaten by many groups.
MEDICINE: Cambium used by the Potawatomi to treat diarrhea, by the Ojibwa for bronchial and heart troubles, and by the Missouri River Indians for bowel troubles. Recent--tannins in bark considered useful as astringent and tonic.
TECHNOLOGY: Bark woven into mats by the Potawatomi.
DYE: Bark used by the Potawatomi for a reddish-brown.
REFERENCES: Densmore 1928:291; Gilmore 1919:75; H. Smith 1928:291; Yanovsky 1936:19.

Quercus shumardii Buckley
Shumard's oak
Native tree; occasional in moist woods in the southern one-third of Illinois.

FOOD: Many similar species were used for food by Historic Indian groups.
REFERENCES: Yanovsky 1936:18-19.

Quercus stellata Wangh.
Post oak
Native tree; occasional to common in the southern one-half of Illinois.

FOOD: Many similar species were used for food by Historic Indian groups.
REFERENCES: Yanovsky 1936:18-19.

Quercus velutina Lam.
Black oak.
Native tree; common in upland woods throughout Illinois.

FOOD: Acorns eaten by the Ojibwa and probably others.
MEDICINE: Bark infusion used for sore eyes by the Menominee and for lung trouble by the Meskwaki. Powdered bark included in dentifrices used by early settlers.
DYE: Inner bark used by the Ojibwa to produce yellow. Recent--used to produce yellow to buff, gold, olive, green and orange.
REFERENCES: Lewis and Elvin-Lewis 1977:245; Lust 1974:552; H. Smith 1923:36, 1928:222, 1932:402,425; Yanovsky 1936:19.

GENTIANACEAE
Gentian Family

Gentiana alba Muhl.
Pale gentian
Native herb; occurs in prairie and on rich wooded slopes throughout Illinois; rare.

SYNONYMS: *G. flavida* Gray
MEDICINE: Root made into a medicinal beverage used as an alterative by the Potawatomi.
REFERENCES: H. Smith 1933:58.

Gentiana andrewsii Griseb.
Closed gentian
Native herb; occasional in moist woods and prairies in the northern one-half of Illinois.

MEDICINE: Root used by the Meskwaki to treat snakebite and caked breasts.
REFERENCES: Lewis and Elvin-Lewis 1977:347; H. Smith 1928:222.

Gentiana puberulenta Pringle
Downy gentian
Native herb; occasional in prairie in the northern two-thirds of Illinois.

SYNONYMS: *G. puberula* Michx., *Dasystephano puberula* (Michx.)Small
MEDICINE: Used as a tonic by the Indians of the Missouri River Region.
REFERENCES: Gilmore 1919:57.

Gentiana quinquefolia L.
Stiff gentian
Native herb; occasional in meadows, prairies, and calcareous woods in the northern two-thirds of Illinois.

MEDICINE: Liquid derived from the root used by the Meskwaki to stop hemorrhage.
REFERENCES: Gilmore 1919:222.

Gentiana saponaria L.
Soapwort gentian
Native herb; occasional in moist, sandy woods in the northeastern portion of Illinois.

MEDICINE: Recent--root used as a bitter tonic.
REFERENCES: Lewis and Elvin-Lewis 1977:375; Tehon 1951:59.

Obolaria virginica L.
Pennywort
Native herb; occurs in rich woods in the southern one-fourth of Illinois; rare.

MEDICINE: Used in a mixture for dressing cuts and wounds by Choctaw.
REFERENCES: Vogel 1970:380.

Sabatia angularis (L.)Pursh
Marsh pink
Native herb; occasional on moist wet soil in the southern two-thirds of Illinois; rare elsewhere.

MEDICINE: Recent--used as a bitter tonic to promote appetite and digestion.
REFERENCES: Lewis and Elvin-Lewis 1977:274; Tehon 1951:100.

Swertia caroliniensis (Walt.)Kuntze
American columbo
Native herb; occasional in rich woods in the southern one-half of Illinois and Cook and Du Page counties.

SYNONYMS: *Frasera caroliniensis* Walt.
MEDICINE: Recent--considered useful as an emetic, cathartic, bitter tonic, and stimulant.
REFERENCES: Lewis and Elvin-Lewis 1977:375; Tehon 1951:56.

GERANIACEAE
Geranium Family

Geranium maculatum L.
Wild geranium
Native herb; common throughout Illinois in rich woods.

MEDICINE: Root used by the Meskwaki and Ojibwa to treat sore gums and pyorrhea and by the Menominee and Ojibwa for dysentery. Recent--considered useful as an astringent; shows experimental hypoglycemic activity.
REFERENCES: Densmore 1928:289; Gilmore 1933:134; Lewis and Elvin-Lewis 1977:219,261,286; H. Smith 1923:36, 1928:222, 1932:370; Vogel 1970:390-91.

HAMAMELIDACEAE
Witch-hazel Family

Hamamelis virginiana L.
Witch-hazel
Native shrub; occasional in woods in the northern one-half of Illinois.

FOOD: Seeds eaten by Indians in the Northwest.
MEDICINE: Infusion of twigs used by the Menominee for an aching back and by the Ojibwa in case of poisoning; twigs and leaves used by the Potawatomi in sweat bath; bark used by some tribes to treat hemorrhoids. Recent--certain compounds have been considered useful for treating internal hemorrhage and as an astringent.
CEREMONY: Seeds used as the sacred beads in the Menominee medicine ceremony.
REFERENCES: Gilmore 1933:131; Lewis and Elvin-Lewis 1977:167,293; H. Smith 1923: 37, 1933:59; Vogel 1970:395-6; Yanovsky 1936:30.

Liquidambar styraciflua L.
Sweet gum
Native tree; occasional to common in low woods in the southern one-third of Illinois.

MEDICINE: Gum used by Indians in the South to treat fevers and as an ointment for sores and wounds. Recent--plant extracts considered useful as stimulant, expectorant, and antiseptic; used for catarrhal affections and as an external ointment; gum used for toothache and cold sores; twigs used by settlers as chewing sticks.
SMOKING: Leaves smoked by the Choctaw.
REFERENCES: Epstein 1981:88; Lewis and Elvin-Lewis 1977:251; Vogel 1970:378.

HIPPOCASTANACEAE
Horse-chestnut Family

Aesculus glabra Willd.
Ohio buckeye
Native tree; occasional in rich woods throughout Illinois.

FOOD: All parts of the tree are poisonous; however, seeds of a similar species were apparently used in California after lengthy processing.
TECHNOLOGY: Used as a fish poison by Indians in the Southeast.
REFERENCES: Lewis and Elvin-Lewis 1977:20,48; Swanton 1946:246; Yanovsky 1936:42.

HYDROPHYLLACEAE
Waterleaf Family

Hydrophyllum appendiculatum Michx.
Great waterleaf
Native herb; occasional in woods throughout Illinois.

FOOD: Young shoots eaten by the Historic Indians of Kentucky.
REFERENCES: Yanovsky 1936:53.

Hydrophyllum canadense L.
Broad-leaf waterleaf
Native herb; occasional to rare in woods throughout Illinois.

FOOD: Roots eaten by Indians in the East during times of famine.
REFERENCES: Yanovsky 1936:53.

Hydrophyllum virginianum L.
Virginia waterleaf
Native herb; occasional to common in woods throughout Illinois.

FOOD: Leaves and young plants eaten as greens by the Iroquois and in Minnesota and Wisconsin.
MEDICINE: Root decoction used to treat flux by the Ojibwa and Menominee.
REFERENCES: H. Smith 1923:37,68, 1932:371; Waugh 1916:117; Yanovsky 1936:53.

HYPERICACEAE
St. John's-wort Family

Hypericum perforatum L.
Common St. John's-wort
Introduced herb; occasional along roadsides and in pastures and fields throughout Illinois.

MEDICINE: This plant has caused livestock poisoning and may cause photodermatitis in some individuals; the flowers contain a volatile oil that is said to help heal cuts and bruises.
DYE: Recent--flowers used for yellow.
REFERENCES: Brooklyn Botanic Garden 1964:23, 1973:40; Lewis and Elvin-Lewis 1977:34,81; Lust 1974:561; Tehon 1951:66.

Hypericum punctatum Lam.
Spotted St. John's-wort
Native herb; occasional in woods and along roadsides throughout Illinois.

MEDICINE: Used by the Meskwaki in a "medicine" mixture.
REFERENCES: H. Smith 1928:223.

Hypericum pyramidatum Ait.
Great St. John's-wort
Native herb; occasional along streambanks in the northern one-half of Illinois.

SYNONYMS: *H. ascyron* L.
MEDICINE: Root used by the Meskwaki and Menominee to treat tuberculosis.
REFERENCES: H. Smith 1928:223.

Triadenum fraseri (Spach.)Gleason
Fraser's St. John's-wort
Native herb; occurs in bogs and wooded swamps in the northern one-half of Illinois; rare.

SYNONYMS: *Hypericum virginicum* (L.)Marsh.
MEDICINE: Leaf decoction used by the Potawatomi to treat fevers.
REFERENCES: H. Smith 1933:60.

JUGLANDACEAE
Walnut Family

Carya cordiformis (Wang.)K.Koch
Bitternut hickory
Native tree; common in woods throughout Illinois.

SYNONYMS: *Hicoria cordiformis* (Wang.)Britton
FOOD: Nuts eaten by Iroquois and probably by other groups as well.
REFERENCES: Yanovsky 1936:16.

Carya glabra (Mill.)Sweet
Pignut hickory
Native tree; occasional to common in woods in the southern one-half of Illinois; rare elsewhere.

SYNONYMS: *Hicoria glabra* (Mill.)Britt., *C. porcina* (Michx.)Nutt.
FOOD: The nuts were eaten by Indians in the eastern states.
REFERENCES: Yanovsky 1936:17.

Carya illinoensis (Wang.)K.Koch
Pecan
Native tree; occasional in bottomland woods in the southern two-thirds of Illinois.

SYNONYMS: *C. olivaeformis* (Michx.)Nutt., *Hicoria pecan* (Marsh.)Britt.
FOOD: The nuts were eaten by Indians throughout the range of the species.
REFERENCES: Yanovsky 1936:17.

Carya laciniosa (Michx.)Loud.
Kingnut hickory
Native tree; occasional to rare in bottomland woods in the southern two-thirds of Illinois.

SYNONYMS: *Carya sulcata* Nutt., *Hicoria laciniosa* (Michx.)Sarg.
FOOD: The large, sweet nuts were eaten by Indians throughout the range of the species.
REFERENCES: Yanovsky 1936:17.

Carya ovalis (Wang.)Sarg.
Sweet pignut hickory
Native tree; occasional to common in woods in the southern one-half of Illinois.

SYNONYMS: *Carya microcarpa* Nutt., *Hicoria microcarpa* (Nutt.)Britt.
FOOD: Nuts eaten by Indians in the eastern states.
REFERENCES: Yanovsky 1936:17.

Carya ovata (Mill.)K.Koch
Shagbark hickory
Native tree; common in woods throughout Illinois.

SYNONYMS: *Juglans squarmosa* Michx., *Hicoria ovata* (Mill.)Britt.
FOOD: Nuts eaten by Indians throughout the range of the species.
TECHNOLOGY: Wood used for bows, arrows, and general utility by the Ojibwa, Potawatomi, Menominee, and undoubtedly others.
REFERENCES: Gilmore 1919:74; H. Smith 1923:68, 1928:259, 1932:405,419, 1933:103,113, 1936:17.

Carya texana Buckl.
Black hickory
Native tree; occasional to rare in upland woods in the southern one- half of Illinois.

FOOD: Nuts of similar species were used for food by Historic Indians.
REFERENCES: Yanovsky 1936:17.

Carya tomentosa (Poir.)Nutt.
Mockernut hickory
Native tree; occasional in woods in the southern three-quarters of Illinois; rare elsewhere.

SYNONYMS: *Hicoria alba* (L.)Britt., *Hicoria tomentosa* Nutt., *Carya alba* (L.)K.Koch.
FOOD: The nuts were eaten by Indians in the South and undoubtedly elsewhere throughout the range of the species.
MEDICINE: Used to treat headache by the Ojibwa.
DYE: Recent--bark used to obtain yellow.
REFERENCES: Densmore 1928:290; Lust 1974:551; Yanovsky 1936:16.

Juglans cinerea L.
Butternut
Native tree; occasional in rich woods throughout Illinois.

FOOD: The nuts were eaten fresh or stored for winter use by Indians throughout the range of the species.
MEDICINE: Sap used as a tonic by the Potawatomi; twigs used as a cathartic by the Meskwaki and as a physic by the Menominee. Recent--leaves and inner bark used as a mild cathartic.
DYE: Husks used for brown by the Ojibwa and Menominee. Recent--green husks used for tan, brown, and gray.
REFERENCES: Brooklyn Botanic Garden 1964:28,32, 1973:20; Densmore 1928:290; Gilmore 1933:127; Lust 1974:553; H. Smith 1923: 68,78, 1928:224,259, 1932:405,425, 1933:103; Tehon 1951:69; Vogel 1970:288-89; Yanovsky 1936:17.

Juglans nigra L.
Black walnut
Native tree; common in rich woods throughout Illinois.

FOOD: Nuts eaten plain or with honey, cooked into soup, or stored for winter use by tribes throughout the range of the species.
MEDICINE: Inner bark used as a physic by the Meskwaki. Recent--leaves and inner bark of root used as a mild cathartic; possible source of antibiotic, cancer chemotherapeutic drug, and mitogenic agent.

DYE: Same as *J. cinerea* L.
REFERENCES: Brooklyn Botanic Garden 1964:29; Gilmore 1919:74, 1933:127; Lewis and Elvin-Lewis 1977:135,362; Lust 1974:552; Parker 1910:99; H. Smith 1928:224,259,271; Tehon 1951:69; Yanovsky 1936:17.

LAMIACEAE
Mint Family

Agastache foeniculum (Pursh)Kuntze
Blue giant hyssop
Native herb; rare on dry soil in Menard County.

SYNONYMS: *A. anethiodora* (Nutt.)Britt.
MEDICINE: Used for cough or chest pain by the Ojibwa.
REFERENCES: Densmore 1928:286.

Agastache scrophulariaefolia (Willd.)Kuntze
Purple giant hyssop
Native herb; occasional in open woods in the northern one-half of Illinois.

MEDICINE: Root decoction used as a diuretic by the Meskwaki.
REFERENCES: H. Smith 1928:225.

Collinsonia canadensis L.
Richweed
Native herb; occasional in rocky woods in the southern tip of Illinois.

MEDICINE: Recent--rootstock and herb considered useful as a diuretic and tonic.
REFERENCES: Lewis and Elvin-Lewis 1977:313; Tehon 1951:42.

Cunila origanoides (L.)Britt.
Dittany
Native herb; occasional to common in dry woods and on sandstone cliffs in the southern one-third of Illinois.

MEDICINE: Recent--considered useful as an aromatic, stimulant, and carminative.
REFERENCES: Tehon 1951:45.

Dracocephalum parviflorum Nutt.
Dragonhead
Introduced herb; occurs in dry soil in the northern one-half of Illinois; uncommon.

SYNONYMS: *Moldavica parviflora* (Nutt.)Britt.
FOOD: Seeds eaten by Historic Indians in Utah and Nevada.
REFERENCES: Yanovsky 1936:57.

Galeopsis tetrahit L.
Common hemp nettle
Introduced herb; occurs on waste ground in the northeastern portion of Illinois; uncommon.

MEDICINE: Plant decoction used for pulmonary trouble by the Potawatomi.
REFERENCES: H. Smith 1933:61.

Glechoma hederacea L.
Ground ivy
Introduced herb; occasional on moist ground throughout Illinois.

SYNONYMS: *Nepeta hederacea* (L.)Trev.
MEDICINE: Recent--used as a stimulant, carminative, tonic, and diuretic. This herb is toxic to livestock.
REFERENCES: Lewis and Elvin-Lewis 1977:56,390; Tehon 1951:83.

Hedeoma hispida Pursh
Rough pennyroyal
Native herb; Occasional in rocky woods and prairies in the northern one-half of Illinois; rare elsewhere.

MEDICINE: Used by the Indians of the Missouri River region as a cold remedy.
REFERENCES: Gilmore 1919:60; Vogel 1970:337-40.

Hedeoma pulegioides (L.)Pers.
American pennyroyal
Native herb; occasional to common in rocky woods, fields, and along roadsides throughout Illinois.

MEDICINE: Herb extract used by the Catawbas as a cold remedy; used by other Indian groups to repel chiggers and for insect bites, to cure flatulence, and for pain, especially headaches. Recent--herb considered useful as an aromatic stimulant and emmenagogue.
REFERENCES: Lewis and Elvin-Lewis 1977:84,169,294,307,345; Tehon 1951:63.

Leonurus cardiaca L.
Motherwort
Introduced herb; occasional in disturbed, shaded areas in the northern one-half of Illinois.

MEDICINE: Recent--considered useful as a stimulant, heart tonic, and diaphoretic; also used for female weakness and hysteria.
DYE: Herb may be used for dark green.
REFERENCES: Lewis and Elvin-Lewis 1977:375; Lust 1974:558; Tehon 1951:72.

Lycopus americanus Muhl.
Common water horehound
Native herb; common on wet ground throughout Illinois.

MEDICINE: Herb used for treatment of stomach cramps by the Meskwaki.
REFERENCES: Lewis and Elvin-Lewis 1977:219; H. Smith 1928:225.

Lycopus asper Greene
Rough water horehound
Native herb; occurs on wet ground in the northern one-fourth of Illinois; rare.

FOOD: Rootstocks dried and boiled for food by Historic Indians in Minnesota and Wisconsin.
REFERENCES: Densmore 1928:320; Yanovsky 1936:54.

Lycopus uniflorus Michx.
Northern bugle weed
Native herb; occasional in marshes and fens and along lakeshores in the northern one-half of Illinois.

FOOD: Rootstocks eaten in British Columbia.
REFERENCES: Yanovsky 1936:54.

Lycopus virginicus L.
Bugle weed
Native herb; occasional on wet ground throughout Illinois.

MEDICINE: Recent--herb considered useful as a sedative, tonic, astringent, and narcotic.
REFERENCES: Tehon 1951:77.

Marrubium vulgare L.
Common horehound
Introduced herb; occasional to common throughout Illinois in fields, pastures and along roadsides.

MEDICINE: Recent--used as a tonic, stimulant, and laxative.
REFERENCES: Tehon 1951:77.

Melissa officinalis L.
Balm
Introduced herb; sometimes cultivated and occasionally escaped.

MEDICINE: Recent--this herb has been used as an aromatic flavoring, diaphoretic, febrifuge, and insect repellent; it is mutagenic.
REFERENCES: Lewis and Elvin-Lewis 1977:94; Tehon 1951:78.

Mentha canadensis L.
Field mint
Native herb; occasional to common in marshes and on low ground throughout Illinois.

SYNONYMS: *M. arvensis* L. var. *villosa* (Benth.)Stewart, *M. arvensis* L. var. *canadensis* L.
FOOD: Pungent leaves used to make a beverage and as a relish in the East.
MEDICINE: Herb used by the Menominee to treat pneumonia, by the Potawatomi for pleurisy and fevers, and by the Ojibwa as a "blood remedy." Recent--used as a carminative.
REFERENCES: H. Smith 1923:39, 1932:371,405, 1933:61; Tehon 1951:79; Vogel 1970:337-40; Yanovsky 1936:54.

Mentha piperita L.
Peppermint
Introduced herb; occasional on moist waste ground throughout Illinois.

MEDICINE: Crushed fresh leaves applied to local pain by Menominee. Recent--leaves and flowers used as aromatic, stimulant, carminative, and antiemetic.
REFERENCES: H. Smith 1923:39, 1932:79; Tehon 1951:79.

Mentha spicata L.
Spearmint
Introduced herb; frequently planted and sometimes escaped.

MEDICINE: Recent--the leaves and flowers have been used as an aromatic, stimulant and carminative; may cause contact dermatitis in some individuals.
REFERENCES: Lewis and Elvin-Lewis 1977:83,294; Tehon 1951:80.

Monarda fistulosa L.
Wild bergamot
Native herb; common throughout Illinois in dry woods, fields, and prairies.

SYNONYMS: *M. mollis* L.
MEDICINE: Herb used by the Menominee and Ojibwa for catarrh; the Meskwaki for colds; the Ojibwa for colds, headaches, skin eruptions, burns, and warts; the Winnebago and Dakotas as a cardiac stimulant; and the Indians of the Missouri River region for abdominal pain. Recent--used as an antifungal, antibacterial, carminative, and anthelminthic.
REFERENCES: Densmore 1928:290; Gilmore 1919:59, 1933:140; Lewis and Elvin-Lewis 1977:194,294,299; H. Smith 1933:39, 1928:225, 1932:372.

Monarda punctata L.
Horsemint
Native herb; occasional in sandy fields and woods, dunes, and prairies in the northern one-half of Illinois.

MEDICINE: Powdered leaves included in a mixture inhaled by the Meskwaki for headache; leaf decoction drunk by the Catawbas for headaches. Recent--used as a diaphoretic, carminative, stimulant, counter-irritant, and vesicant.
REFERENCES: Lewis and Elvin-Lewis 1977:167,280,294; H. Smith 1928:225; Tehon 1951:81; Vogel 1970:337-40.

Nepeta cataria L.
Catnip
Introduced herb; occasional to common in open woods, fields, and waste ground throughout Illinois.

MEDICINE: Herb used by Menominee in a treatment for pneumonia and by Ojibwa as a blood purifier and for fevers. Recent--used in Ozarks as a toothache remedy, cold remedy, for infantile colic, and as a stimulant; possibly hallucinogenic.
REFERENCES: Densmore 1928:290; Lewis and Elvin-Lewis 1977:257,276,307; H. Smith 1923:39, 1932:372; Tehon 1951:82.

Physostegia parviflora Nutt.
False dragonhead
Native herb; occurs in moist prairies in the northeastern portion of Illinois; rare.

MEDICINE: Leaf decoction used by the Meskwaki to treat colds.
REFERENCES: H. Smith 1928:226.

Prunella vulgaris L.
Self-heal
Introduced and native herb; occasional in lawns, fields, and waste ground throughout Illinois.

MEDICINE: Herb used for "female troubles" by the Ojibwa.
CEREMONY: Herb used by the Ojibwa to sharpen powers of observation while hunting.
REFERENCES: H. Smith 1932:372,430; Tehon 1951:91.

Pycnanthemum virginianum (L.)Dur. & Jacks.
Common mountain mint
Native herb; occasional to common in marshes, fens, and prairies in the northern one-half of Illinois; local elsewhere.

FOOD: Flowers and buds used for seasoning meat and broth by the Ojibwa.
MEDICINE: Herb used for "female troubles" by the Ojibwa and infusion used by Meskwaki for chills and ague; used in the Southeast for diarrhea and other bowel complaints.
TECHNOLOGY: Herb used by the Meskwaki to scent mink traps and chewed tops used in catching snakes.
REFERENCES: Densmore 1928:290; Lewis and Elvin-Lewis 1977:286; H. Smith 1928:226,273; Yanovsky 1936:54.

Salvia lyrata L.
Cancer-weed
Native herb; occasional in rich open woods in the southern one-sixth of Illinois.

MEDICINE: Root chewed by the Catawbas to make a salve for sores.
REFERENCES: Vogel 1970:359.

Scutellaria epilobifolia Muhl.
Marsh skullcap
Native herb; occasional in marshes in the northern one-half of Illinois.

SYNONYMS: *S. galericulata* L.
MEDICINE: Used by the Ojibwa for heart trouble.
REFERENCES: H. Smith 1932:372.

Scutellaria lateriflora L.
Mad-dog skullcap
Native herb; occasional to common throughout Illinois in marshes and swampy woods and along the borders of streams.

MEDICINE: Medicinal beverage used by the Cherokee to promote menstruation and to treat diarrhea. Used by early Euro-American settlers in treating hydrophobia; more recently, as a tonic, nervine, and antispasmodic.
REFERENCES: Vogel 1970:366-67; Tehon 1951:104.

Scutellaria parvula Michx.
Small skullcap
Native herb; occasional throughout Illinois in rocky woods, prairies, fields, and limestone barrens.

MEDICINE: Used by the Meskwaki to treat diarrhea.
REFERENCES: H. Smith 1928:227; Vogel 1970:367.

Stachys hispida Pursh
Hairy hedge nettle
Native herb; occasional in low woods, swamps, and marshes in the northern one-half of Illinois; rare elsewhere.

SYNONYMS: *S. tenuifolia* var. *hispida* (Pursh)Fern.
MEDICINE: Leaves used by the Meskwaki to treat a bad cold.
REFERENCES: H. Smith 1928:227.

Stachys palustris L.
Woundwort
Native herb; occasional to common in wet prairies, swampy, or marshy ground in the northern one-half of Illinois; rare elsewhere.

MEDICINE: Used by the Ojibwa to treat colic.
REFERENCES: Densmore 1928:293.

Teucrium canadense L.
American germander
Native herb; common throughout Illinois in low woods and wet prairies.

MEDICINE: Recent--herb used as an aromatic, stimulant, diaphoretic, diuretic, and emmenagogue.
REFERENCES: Tehon 1951:112.

LAURACEAE
Laurel Family

Lindera benzoin (L.)Blume
Spicebush
Native shrub; common throughout Illinois in rich woods.

SYNONYMS: *Benzoin aestivale* (L.)Nees
FOOD: Leaves used by the Ojibwa to make a tealike beverage and as a flavoring.
MEDICINE: Used by the Ojibwa in "medicine."
REFERENCES: Gilmore 1933:131; Yanovsky 1936:26.

Sassafras albidum (Nutt.)Nees
Sassafras
Native tree; common in rich woods in the southern three-fourths of Illinois; rare elsewhere.

SYNONYMS: *S. variifolium* (Salisb.)Kuntze, *S. officinale* Nees & Eberm., *Laurus sassafras* L., *S. sassafras* (L.)Kart.
FOOD: Leaves used in soups, as flavoring, and to make a beverage by Indians in southern and eastern states. Recent--leaves important as a flavoring and thickening agent in Creole cooking.
MEDICINE: Used by the Meskwaki for internal troubles and by the Ojibwa as a blood thinner. Recent--bark of root and twigs used as an aromatic stimulant and flavoring agent, twigs used as chewing sticks, and in commercial dental poultices. Safrole is carcinogenic and hepatotoxic, possibly hallucinogenic in large doses, and may cause contact dermatitis in some individuals.
DYE: Recent--flowers may be used for yellow and roots for brown.
SMOKING: Leaves apparently smoked by some Indian groups in the South.
REFERENCES: Brooklyn Botanic Garden 1964:94, 1973:6; Epstein 1981:131; Gilmore 1933:130; Lewis and Elvin-Lewis 1977:84,121,231,250; Lust 1974:561; H. Smith 1928:194; Tehon 1951:103; Yanovsky 1936:26.

LEGUMINOSAE
Pea Family

Amorpha canescens Pursh
Leadplant
Native shrub; common in prairies in the northern two-thirds of Illinois; local elsewhere.

MEDICINE: Leaves used by the Meskwaki as a vermifuge and to treat eczema.
SMOKING: Leaves smoked by Indians of the Missouri River region.
REFERENCES: Epstein 1981:91; H. Smith 1928:227; Vogel 1970:183.

Amorpha fruticosa L.
False indigo
Native shrub; occasional throughout Illinois on moist soil.

MEDICINE: Used in medicine by the Meskwaki.
REFERENCES: H. Smith 1928:227.

Amphicarpa bracteata (L.)Fern.
Hog peanut
Native herb; occasional throughout Illinois in woods and thickets.

SYNONYMS: *Falcata comosa* (L.)Kuntze, *A. monoica* (L.)Ell., *A. bracteata* (L.)Fern. var. *comosa* (L.)Fern., *A. comosa* (L.)G.Don.
FOOD: Underground fruits eaten raw or boiled by Indians in the eastern and central states.
MEDICINE: Used as a physic by the Ojibwa.
REFERENCES: Densmore 1928:289,320; Gilmore 1919:95; H. Smith 1928:259; Yanovsky 1936:37.

Apios americana Medic.
Groundnut
Native herb; common throughout Illinois in woods and thickets.

SYNONYMS: *Glycine apios* L., *A. tuberosa* Moench.
FOOD: Tubers eaten raw, boiled, or roasted by Indians in the eastern and central states (sometimes with maple sugar).
REFERENCES: Gilmore 1919:94; Parker 1910:105; H. Smith 1923:68, 1928:260, 1933:103; Waugh 1916:120; Yanovsky 1936:37.

Apios priceana Robins.
Price's groundnut
Native herb; occurs in woods in Union County; very rare.

FOOD: The smaller tubers of the similar *Apios americana* were used for food by many tribes.
REFERENCES: Yanovsky 1936:37.

Astragalus canadensis L.
Canadian milk vetch
Native herb; occasional throughout Illinois in prairies and thickets.

SYNONYMS: *A. carolinianus* L.
FOOD: The roots were gathered by Blackfeet in spring and fall and eaten either raw or boiled. (Some related species are toxic.)
REFERENCES: Yanovsky 1936:36.

Baptisia leucantha T.&G.
White wild indigo
Native herb; occasional throughout Illinois in prairies and woods.

MEDICINE: Boiled roots used by the Meskwaki for catarrh, used elsewhere for wounds, sores, eczema, typhus, dysentery and snakebite. The plant is toxic to livestock.
DYE: Recent--the leaves and fruit can be used for blue.
REFERENCES: Lewis and Elvin-Lewis 1977:42; H. Smith 1928:228; Vogel 1970:323; Voigt and Mohlenbrock [1979]:72.

Baptisia leucophaea Nutt.
Cream wild indigo
Native herb; occasional in moist prairies and open woods in the northern four-fifths of Illinois.

DYE: Recent--the leaves and fruit can be used to make a somewhat inferior indigo blue. This species is toxic to livestock.
REFERENCES: Lewis and Elvin-Lewis 1977:42; Voigt and Mohlenbrock [1979]:74.

Cassia marilandica L.
Maryland senna
Native herb; occasional throughout Illinois along roadsides and in thickets.

MEDICINE: Seeds used by the Meskwaki to treat sore throats. Recent--considered useful as a cathartic and purgative. In large quantities, this species is toxic to livestock.
REFERENCES: H. Smith 1928:228; Tehon 1951:33; Vogel 1970:366.

Cercis canadensis L.
Redbud
Native tree; occasional to common in woods throughout Illinois.

FOOD: The pods of a similar species were eaten by the Navajo.
REFERENCES: Yanovsky 1936:36.

Desmanthus illinoensis (Michx.)MacM.
Illinois mimosa
Native herb; occasional in prairies and along levees in the south and north; uncommon elsewhere.

SYNONYMS: *Acuan illinoensis* (Michx.)Kuntze
MEDICINE: Used by the Indians of the Missouri River region to treat itching.
REFERENCES: Gilmore 1919:37.

Desmodium illinoense Gray
Illinois tick trefoil
Native herb; occasional throughout Illinois in prairies and along roadsides.

TECHNOLOGY: Root juice used by the Meskwaki and Potawatomi to lure wild animals into traps.
REFERENCES: H. Smith 1928:228.

Gleditsia triacanthos L.
Honey locust
Native tree; common in woods in the southern portion of Illinois, local elsewhere.

FOOD: Beer was made by fermenting sweet pulp in pods by the Indians in the Mississippi River region.
MEDICINE: Bark tea was used to treat measles, fever, smallpox, and head colds by the Meskwaki.
REFERENCES: H. Smith 1928:228; Yanovsky 1936:36.

Glycyrrhiza lepidota Pursh
Wild licorice
Introduced herb; occasional on waste ground in the northeastern portion of Illinois.

MEDICINAL USAGE: Used by the Indians of the Missouri River region for toothache, headache, and fevers.
REFERENCES: Gilmore 1919:40; Vogel 1970:200,322.

Gymnocladus dioica (L.)K.Koch
Kentucky coffee-tree
Native tree; common throughout Illinois in low woods.

SYNONYMS: *G. canadensis* Lam.
FOOD: Seeds roasted and eaten like nuts by the Meskwaki or ground and used to make a beverage. However, the seeds and pulp are poisonous.
MEDICINE: Wax from pods used by the Meskwaki to treat insanity and by the Indians of the Missouri River region to treat constipation.
REFERENCES: Gilmore 1919:37; H. Smith 1928:22,260; Vogel 1970:185; Yanovsky 1936:36.

Lathyrus maritimus (L.)Bigel.
Beach pea
Native herb; occurs on beaches near Lake Michigan; rare.

FOOD: Fresh stalks and sprouts eaten raw or cooked by the Iroquois.
REFERENCES: Yanovsky 1936:37.

Lathyrus ochroleucus Hook.
Marsh vetchling
Native herb; occasional in open woods and on moist ground in the northern one-half of Illinois.

FOOD: Seeds eaten by the Ojibwa.
TECHNOLOGY: Root used by the Meskwaki to lure beaver and other game into traps.
REFERENCES: H. Smith 1928:273; Yanovsky 1936:37.

Lathyrus palustris L.
Marsh vetchling
Native herb; occasional in the northern one-half of Illinois in open woods and on moist ground.

FOOD: Seeds eaten by the Ojibwa.
TECHNOLOGY: Root used by the Meskwaki to lure beaver and other game into traps.
REFERENCES: H. Smith 1928:273; Yanovsky 1936:37.

Lathyrus venosus Muhl.
Veiny pea
Native herb; occasional in dry woods and prairies in the northern one- fourth of Illinois.

FOOD: Seeds used as food by the Ojibwa.
MEDICINE: Root decoction used by the Ojibwa to treat convulsions and hemorrhaging.
CEREMONY: Used as a charm by the Ojibwa.
REFERENCES: Densmore 1928:290.

Medicago sativa L.
Alfalfa
Introduced herb; commonly cultivated and often escaped.

FOOD: Recent--sprouted seeds sometimes eaten.
MEDICINE: The herb contains a relatively high amount of Vitamin K.
DYE: Recent--seed can be used for yellow.
REFERENCES: Tehon 1951:77.

Melilotus officinalis (L.)Lam.
Yellow sweet clover
Introduced herb; common throughout Illinois on waste ground.

MEDICINE: Recent--the herb has been used as an expectorant, diuretic, anticoagulant, and source of the antibiotic dicumarol.
REFERENCES: Lewis and Elvin-Lewis 1977:192; Tehon 1951:78.

Petalostemum candidum (Willd.)Michx.
White prairie clover
Native herb; occasional throughout Illinois on sandy or gravelly soil, in prairies, or in open woods.

FOOD: Root chewed or eaten raw and the leaves made into a tealike beverage by Nebraska Indians.
REFERENCES: Yanovsky 1936:38.

Petalostemum purpureum (Vent.)Rydb.
Purple prairie clover
Native herb; occasional throughout Illinois in prairies.

FOOD: Roots chewed and the leaves made into a tealike beverage by Indians of the Missouri River region.
MEDICINE: Root decoction used by the Meskwaki for measles and diarrhea and by the Ojibwa to treat heart trouble.
TECHNOLOGY: Used by the Indians of the Missouri River region to make brooms and as a prophylactic.
REFERENCES: Densmore 1928:291; Gilmore 1919:42,94; H. Smith 1928:229,230; Yanovsky 1936:38.

Psoralea tenuiflora Pursh
Scurf-pea
Native herb; occasional in prairies in the northern two-thirds of Illinois.

MEDICINE: Used by the Indians of the Missouri River region to treat tuberculosis.
REFERENCES: Gilmore 1919:41.

Robinia pseudoacacia L.
Black locust
Native tree; common in woods and thickets throughout Illinois although only native to southeastern Illinois.

FOOD: Seeds boiled for food in Pennsylvania.
MEDICINE: Bark used as a seasoner of medicine by the Menominee. Inner bark, young leaves, and seeds are poisonous and lectinic.
REFERENCES: Lewis and Elvin-Lewis 1977:45; H. Smith 1923:40; Vogel 1970:107; Yanovsky 1936:39.

Strophostyles helvola (L.)Britt.
Wild bean
Native herb; occasional throughout Illinois in rocky woods, sandbars, roadsides, and fields.

SYNONYMS: *Phaseolus diversifolius* Pers.
FOOD: Boiled tubers eaten by Indians in Louisiana and probably elsewhere.
REFERENCES: Yanovsky 1936:38.

Tephrosia virginiana (L.)Pers.
Goat's-rue
Native herb; occasional throughout Illinois in dry prairies and woods.

MEDICINE: Herb tea used by Plains Indians for worms, by the Cherokee for lassitude, and by the Creeks for chronic coughing and bladder trouble. Recent--used as a cathartic, tonic, vermifuge, insecticide, and fish poison.
REFERENCES: Tehon 1951:111; Vogel 1970:299.

Trifolium pratense L.
Red clover
Introduced herb; common throughout Illinois on waste ground.

MEDICINE: Recent--the herb has been used as an alterative and sedative.
REFERENCES: Tehon 1951:113.

Vicia caroliniana Walt.
Wood vetch
Native herb; occasional on dry, wooded slopes in the northern one- fourth of Illinois.

MEDICINE: Used as a "medicine" by the Cherokee.
REFERENCES: Vogel 1970:102.

LOGANIACEAE
Logania Family

Spigelia marilandica L.
Indian pink
Native herb; occasional in woods in the southern one-fourth of Illinois.

MEDICINE: Herb or root used as a vermifuge by the Cherokee. Recent--also used as a vermifuge.
REFERENCES: Lewis and Elvin-Lewis 1977:291; Vogel 1970:348-49.

LORANTHACEAE
Mistletoe Family

Phoradendron flavescens (Pursh)Nutt.
Mistletoe
Native herb; occasional in the southern one-sixth of Illinois, parasitic on various deciduous trees.

MEDICINE: Leaf tea used as a contraceptive by California Indians and by some North American Indians for stopping post-partum hemorrhage. Recent--considered useful as an emmenagogue and hemostatic. The berries are toxic in large quantities.
REFERENCES: Lewis and Elvin-Lewis 1977:48-49,190,323; Tehon 1951:86; Vogel 1970:244.

MAGNOLIACEAE
Magnolia Family

Liriodendron tulipifera L.
Tulip tree
Native tree; common in rich woods in the southern two-thirds of Illinois.

MEDICINE: Crushed leaves used by southern Indians as a poultice for headache, fruit and bark for ague, and root for toothache; used by the Catawbas as a vermifuge. Recent--bark considered useful as a tonic, aromatic, and diaphoretic.
REFERENCES: Lewis and Elvin-Lewis 1977:250; Tehon 1951:74; Vogel 1970:385-86.

Magnolia acuminata L.
Cucumber magnolia
Native tree; occasional in rich woods in the southern tip of Illinois.

MEDICINE: Fruit and bark used by Indians of the lower Mississippi region as a vermifuge and to treat sores and fever. Recent--used as an aromatic and bitter tonic.
REFERENCES: Tehon 1951:76; Vogel 1970:332-34.

MALVACEAE
Mallow Family

Althaea rosea (L.)Cov.
Hollyhock
Introduced herb; cultivated and occasionally escaped.

DYE: Recent--the flowers can be used for blue.
REFERENCES: Lust 1974:556.

Callirhoe involucrata (T.&G.)Gray
Poppy mallow
Introduced herb; occasional on cultivated ground and along roadsides.

MEDICINE: Used by the Indians of the Missouri River region to treat head colds, aches, and internal pains.
REFERENCES: Gilmore 1919:51.

Gossypium hirsutum L.
Cotton
Introduced herb; cultivated and sometimes escaped.

DYE: Recent--petals sometimes used for brown and yellow.
REFERENCES: Lust 1974:554.

Malva rotundifolia L.
Round-leaved mallow
Introduced herb; occasional throughout Illinois on waste ground.

MEDICINE: Recent--herb and root used as an emollient and demulcent.
REFERENCES: Tehon 1951:76.

Napaea dioica L.
Glade mallow
Native herb; occasional on alluvial soil in the northern one-half of Illinois.

MEDICINE: Root used by the Meskwaki to treat hemorrhoids.
REFERENCES: H. Smith 1928:232.

MARTYNIACEAE
Martynia Family

Proboscidea louisianica (Mill.)Thell.
Unicorn-plant
Introduced herb; occasional throughout Illinois on river banks and waste ground.

SYNONYMS: *Martynia louisiana* Mill., *M. proboscidea* Glax.
FOOD: Young pods eaten by the Apaches.
TECHNOLOGY: Pod fibers used in basketry by Indians in the Southwest.
REFERENCES: Yanovsky 1936:57.

MENISPERMACEAE
Moonseed Family

Cocculus carolinus (L.)DC.
Snailseed
Native vine; occasional on streambanks in southern Illinois.

TECHNOLOGY: Used as a fish poison by southern Indians.
REFERENCES: Swanton 1946:246.

Menispermum canadense L.
Moonseed
Native vine; common throughout Illinois in moist woods and thickets.

MEDICINE: Root used for "medicine" by the Ojibwa. Recent--rhizome and roots considered useful as a tonic, alterative, and diuretic. The fruit is poisonous.
REFERENCES: Lewis and Elvin-Lewis 1977:29; H. Smith 1932:375; Tehon 1951:78; Vogel 1970:289.

MENYANTHACEAE
Buckbean Family

Menyanthes trifoliata L.
Buckbean
Native herb; occurs in bogs and marshes in the northeastern portion of Illinois; uncommon.

MEDICINE: Used in some manner as a "medicine" by the Menominee. Recent--used as a tonic and laxative.
REFERENCES: H. Smith 1923:36; Tehon 1951:80; Vogel 1970:286.

MORACEAE
Mulberry Family

Maclura pomifera (Raf.)Schneider
Osage orange
Introduced tree; occasional in woods and fence rows throughout Illinois.

MEDICINE: Root decoction used by the Comanches for sore eyes.
TECHNOLOGY: Wood valued by southern Indians for bows.
DYE: Recent--bark may be used for yellow and gold.
REFERENCES: Brooklyn Botanic Garden 1964:27; Lewis and Elvin-Lewis 1977:97,224; Lust 1974;559; Swanton 1946:578.

Morus alba L.
White mulberry
Introduced tree; occasional in woods throughout Illinois.

DYE: Recent--bark and wood may be used for yellow.
REFERENCES: Brooklyn Botanic Garden 1964:58.

Morus rubra L.
Red mulberry
Native tree; common throughout Illinois in woods.

FOOD: Fruit eaten by Indians throughout range of the species.
MEDICINE: Root bark used by southern Indians to treat "any sickness"; sap used by Rappahannocks to cure ringworm.
REFERENCES: H. Smith 1928:251; Swanton 1946:245,578; Vogel 1970:220; Waugh 1916:128; Yanovsky 1936:20.

MYRICACEAE
Bayberry Family

Comptonia peregrina (L.)Coult.
Sweet-fern
Native shrub; occurs on sand flats and in barrens in the northern one-fourth of Illinois; uncommon.

SYNONYMS: *Myrica asplenifolia* H.&G.
FOOD: Tealike beverage prepared from leaves by the Ojibwa, Potawatomi, and Menominee.

MEDICINE: Leaf tea used by the Ojibwa for diarrhea and stomach cramps, by the Potawatomi for itching, and by Menominee to hasten parturition; boiled leaves used by some groups as a poultice for toothache. Recent--used as an astringent and stimulant.
CEREMONY: Leaves used for ceremonial incense by the Ojibwa.
REFERENCES: Gilmore 1919:127; Lewis and Elvin-Lewis 1977:255,287,322; H. Smith 1923:42, 1932:375, 1933:65; Tehon 1951:82; Vogel 1970:247; Yanovsky 1936:16.

NELUMBONACEAE
Lotus Family

Nelumbo lutea (Willd.)Pers.
American lotus
Native herb; occasional throughout Illinois in lakes and ponds.

FOOD: Tubers boiled, roasted, or dried; seeds shelled and cooked; young leaves used as greens by Indians throughout the distribution of the species. This plant was also introduced by the Indians into lakes and ponds outside of its original range.
MEDICINE: Used by the Meskwaki as a nourishing convalescent.
REFERENCES: Gilmore 1919:79; H. Smith 1928:194,262, 1932:407, 1933:105.

NYCTAGINACEAE
Four-o'clock Family

Mirabilis nyctaginea (Michx.)MacM.
Wild four-o'clock
Introduced herb; common throughout Illinois in waste areas.

SYNONYMS: *Oxybaphus nyctaginea* (Michx.)Sweet, *Allionia nyctaginea* Michx.
MEDICINE: Whole herb or root used in treating bladder problems by the Meskwaki, by the Ojibwa to reduce swellings, and by the Indians of the Missouri River region for fever, worms, and swellings.
REFERENCES: Densmore 1928:294; Gilmore 1919:26; H. Smith 1928:232, 1932:375.

NYMPHAEACEAE
Water-lily Family

Nuphar luteum L. ssp. *macrophyllum* (Small)Beal
Yellow pond-lily
Native herb; occasional throughout Illinois in ponds and slow streams.

SYNONYMS: *Nymphaea advena* Ait., *Nuphar advena* (Ait.)Ait.f.
FOOD: Thick tubers eaten by Indians throughout most of the range of the species; identified as *macoupin* by Trowbridge (1938).
MEDICINE: Powdered root used as a poultice for cuts and swellings by the Ojibwa, Menominee, and Potawatomi; grated root used for sores by the Ojibwa; root used as a convalescent and the seeds used for sore throats by the Meskwaki.
REFERENCES: Gilmore 1919:79; Parker 1910:105; H. Smith 1923:43,69, 1928:194,195,198, 1932:376, 1933:65; Yanovsky 1936:25.

Nymphaea odorata Ait.
Fragrant water-lily
Native herb; occurs in lakes and shallow ponds; rare.

SYNONYMS: *Castalia odorata* (Ait.)Woodville and Wood
FOOD: Flower buds eaten by the Ojibwa.
MEDICINE: Root used by the Ojibwa to prepare a cough medicine for tuberculosis and for sore mouths, by the Meskwaki for eczema, and by the Potawatomi as a poultice. Recent--used as an astringent.
REFERENCES: Densmore 1928:288; H. Smith 1923:42, 1928:194, 1932:376,407, 1933:65; Tehon 1951:34; Vogel 1970:249; Yanovsky 1936:25.

Nymphaea tuberosa Paine
White water-lily
Native herb; occasional in ponds and streams in the northern two-thirds of Illinois.

SYNONYMS: *Castalia tuberosa* (Paine)Greene
FOOD: Tubers, leaf and flower buds, and seeds are supposedly edible, but there is no recorded use by historic Indians.
REFERENCES: Yanovsky 1936:25.

NYSSACEAE
Sour Gum Family

Nyssa sylvatica Marsh.
Sour gum
Native tree; occasional in bogs and swamps in the southern one-third of Illinois.

TECHNOLOGY: Wood used by the Ojibwa for awl handles, mauls, and war clubs.
DYE: Recent--bark can be used for brown.
REFERENCES: Gilmore 1919:138; Lust 1974:552.

OLEACEAE
Ash Family

Fraxinus americana L.
White ash
Native herb; common throughout Illinois in woods.

MEDICINE: Bark infusion used by the Meskwaki to cure sores and itches. Recent--the inner bark has been used as a tonic, cathartic, diuretic, and astringent for hemorrhoids.
TECHNOLOGY: Wood used by the Ojibwa for fish spears and in canoe and snowshoe making.
REFERENCES: Densmore 1928:289; Gilmore 1933:139-40; Lewis and Elvin-Lewis 1977:294,352; H. Smith 1928:233; Tehon 1951:57; Vogel 1970:275-76.

Fraxinus nigra Marsh.
Black ash
Native tree; occasional in moist woods and on slopes in the northern one-half of Illinois.

MEDICINE: Used in some manner by the Menominee and Ojibwa for "medicine"; infusion of inner bark used by the Meskwaki for internal ailments. More recent use same as *F. americana*.
TECHNOLOGY: Wood used by the Ojibwa for basketry splints, bows, and arrows.
DYE: Bark used by the Ojibwa for blue.
REFERENCES: Densmore 1928:289; Gilmore 1933:139; H. Smith 1923:43, 1928:233, 1932:420; Tehon 1951:57; Vogel 1970:276.

Fraxinus pennsylvanica Marsh.
Red ash
Native tree; common throughout Illinois in moist woods.

FOOD: Cambium sometimes cooked for food by the Ojibwa.
MEDICINE: Inner bark used by the Ojibwa as a tonic.
TECHNOLOGY: Wood used by the Ojibwa, Potawatomi, and Indians of the Missouri River region for basketry splints, spoons, bows, arrows, snowshoes, sleds, cradleboards, etc.
REFERENCES: Densmore 1928:289; Gilmore 1919:56; H. Smith 1932:376,407,420, 1933:66,113; Vogel 1970:276; Yanovsky 1936:52.

ONAGRACEAE
Evening-primrose Family

Epilobium adenocaulon Haussk.
Northern willow-herb
Native herb; occasional in bogs and calcareous wet areas in the northern one- fourth of Illinois.

MEDICINE: Root tea used to treat diarrhea by the Potawatomi.
REFERENCES: H. Smith 1933:66.

Epilobium angustifolium L.
Fireweed
Native herb; local in the northern one-fifth of Illinois on dunes and bogs after a fire.

FOOD: Gelatinous contents of stalks eaten by Indians in the Northeast.
MEDICINE: Root used by the Ojibwa as a poultice for boils, by the Menominee and Ojibwa for swellings, and by the Potawatomi for "medicine". Recent--leaves and roots have been used as a tonic, astringent, demulcent, and emollient.
REFERENCES: Densmore 1928:289; Lewis and Elvin-Lewis 1977:391; H. Smith 1923:43, 1932:376, 1933:66; Tehon 1951:51; Yanovsky 1936:46.

Oenothera biennis L.
Evening primrose
Native herb; common throughout Illinois in fields, prairies, and waste ground.

MEDICINE: Plant used by the Ojibwa to make poultices for bruises and by the Potawatomi for "medicine."
REFERENCES: H. Smith 1932:376, 1933:66.

OROBANCHACEAE
Broomrape Family

Orobanche fasciculata Nutt.
Clustered broomrape
Native herb; parasitic on *Artemisia* and other composites growing on sandy soil in the northern one-half of Illinois; uncommon.

FOOD: The plant was eaten by Indians in Utah and Nevada.
REFERENCES: Yanovsky 1936:57.

Orobanche ludoviciana Nutt.
Broomrape
Native herb; parasitic on various plants, particularly *Artemisia*, *Ambrosia*, and other composites growing in sandy soil; rare.

FOOD: The rootstocks were eaten by Indians in the West.
REFERENCES: Yanovsky 1936:57.

OXALIDACEAE
Wood-sorrel Family

Oxalis corniculata L.
Creeping wood sorrel
Introduced herb; occasional throughout Illinois on waste ground.

FOOD: This plant is toxic in large quantities.
DYE: Plant used by the Menominee for yellow.
REFERENCES: Lewis and Elvin-Lewis 1977:47; H. Smith 1923:78.

Oxalis stricta L.
Yellow wood sorrel
Native herb; common throughout Illinois in woods, fields, and along roadsides.

FOOD: The plant eaten by Indians in Nebraska. This plant is toxic in large quantities.
DYE: Root used by the Meskwaki and plant by the Menominee for yellow.
REFERENCES: Gilmore 1919:98; Lewis and Elvin-Lewis 1977:47; H. Smith 1923:78, 1928:271; Yanovsky 1936:40.

Oxalis violacea L.
Violet wood sorrel
Native herb; common throughout Illinois in woods, prairies, and on bluffs.

FOOD: Leaves, flowers, and bulbs eaten by Indian children in Nebraska. This plant is toxic in large quantities.
REFERENCES: Gilmore 1919:98; Yanovsky 1936:40.

PAPAVERACEAE
Poppy Family

Corydalis aurea Willd.
Golden corydalis
Native herb; occurs on rocky soil in the northeastern portion of Illinois; rare.

MEDICINE: Root used by the Ojibwa to clear the head and as a reviver.
REFERENCES: H. Smith 1932:370.

Dicentra canadensis (Goldie)Walp.
Squirrel-corn
Native herb; occasional throughout Illinois in rich woods.

FOOD: Tubers eaten by Indians in New York state.
MEDICINE: Recent--the tubers have been used as a tonic, diuretic, and alterative; the plants are toxic to livestock in large quantities.
REFERENCES: Lewis and Elvin-Lewis 1977:375; Tehon 1951:48; Vogel 1970:295; Yanovsky 1936:26.

Dicentra cucullaria (L.)Bernh.
Dutchman's breeches
Native herb; common throughout Illinois in rich woods.

MEDICINE: Recent--the tubers have been used as a tonic, diuretic, and alterative; the plants are toxic to livestock in large quantities.
CEREMONY: Root was used by the Menominee as a love charm.
REFERENCES: Lewis and Elvin-Lewis 1977:32; H. Smith 1923:81; Tehon 1951:48; Vogel 1970:295.

Fumaria officinalis L.
Fumitory
Introduced herb; cultivated and occasionally escaped.

MEDICINE: Recent--the herb has been used as a tonic and blood purifier.
DYE: Recent--the herb can be used for yellow and green.
REFERENCES: Lewis and Elvin-Lewis 1977:375; Lust 1974:555.

Sanguinaria canadensis L.
Bloodroot
Native herb; common throughout Illinois in rich woods.

MEDICINE: Root used by the Menominee to strengthen the effect of other medicines, by the Ojibwa for sore throats, by the Meskwaki for burns, by the Potawatomi for diphtheria, by the Indians of the Missouri River region for rheumatism, and by some groups for toothache. Recent--the roots have been used as an expectorant, sternutatory, and emetic.
DYE: The rootstocks were used by the Ojibwa, Menominee, Potawatomi, and Indians of the Missouri River region for red or orange dye or facepaint.
CEREMONY: Used as a love charm by the Indians of the Missouri River region.
REFERENCES: Densmore 1928:292; Gilmore 1919:31, 1933:31; Lewis and Elvin-Lewis 1977:135, 251,278; H. Smith 1923:44,78, 1928:234,271, 1932:377,426, 1933:68,121; Tehon 1951:101; Vogel 1970:354,356.

PASSIFLORACEAE
Passion-flower Family

Passiflora incarnata L.
Passion-flower
Native vine; occasional on dry soil in the southern one-third of Illinois.

FOOD: Fruit eaten by Indians in the southern states.
MEDICINE: Recent--the flowers, fruits, and roots have been used as a uterine sedative and hemostatic as well as for insomnia, convulsions, and spasmodic disorders, neuralgia, and epilepsy.
REFERENCES: Lewis and Elvin-Lewis 1977:168; Tehon 1951:85; Yanovsky 1936:43.

Passiflora lutea L.
Passion-flower
Native vine; occasional to common in woods and thickets in the southern one-half of Illinois.

MEDICINE: Same uses as *P. incarnata*.
REFERENCES:Tehon 1951:85.

PHRYMACEAE
Lopseed Family

Phryma leptostachya L.
Lopseed
Native herb; occasional throughout Illinois in rich woods.

MEDICINE: Root used by the Ojibwa for sore throats.
REFERENCES: Densmore 1928:291.

PHYTOLACCACEAE
Pokeweed Family

Phytolacca americana L.
Pokeweed
Native herb; occasional to common throughout Illinois in woods and waste ground.

FOOD: Young leaves and stalks eaten by the Iroquois and undoubtedly other Indian groups. This plant is often collected as a potherb even today. However, the entire plant, especially the roots and seeds, are poisonous and mitogenic and even skin contact should be avoided.
MEDICINE: Berries used for rheumatism by the Indians of Virginia. Recent--the berries, roots, and leaves have been used as an alterative, emetic, and purgative.
DYE: The fruit was used by the Indians of the Missouri River region as a red paint although it fades fairly rapidly.
REFERENCES: Gilmore 1919:27; Lewis and Elvin-Lewis 1977:32,90,97,167; Vogel 1970:350-51; Yanovsky 1936:23.

PLANTAGINACEAE
Plantain Family

Plantago lanceolata L.
Buckhorn plantain
Introduced herb; common throughout Illinois on waste ground.

MEDICINE: Recent--the leaves and roots have been used as an astringent and alterative.
REFERENCES: Tehon 1951:88.

Plantago major L.
Common plantain
Introduced herb; common on waste ground in the northern one-third of Illinois, local elsewhere.

MEDICINE: Used by the Meskwaki and Menominee for burns, swellings, and urinary troubles, by the Potawatomi to stop choking and for swellings and infections, and by the Indians of the Missouri River region for splinters. Recent--leaves used as a tonic and astringent.
CEREMONY: Used to ward off snakes by the Ojibwa.
REFERENCES: Densmore 1928:291,263; H. Smith 1928:234, 1932:380,431, 1933:71; Tehon 1951:88.

Plantago rugelii Dcne.
Rugel's plantain
Native herb; common throughout Illinois in fields, woods, and waste ground.

MEDICINE: Leaves used by the Menominee as a poultice to reduce swellings.
REFERENCES: H. Smith 1923:46.

PLATANACEAE
Sycamore Family

Platanus occidentalis L.
Sycamore
Native tree; common in moist woods and along streams in the southern two-thirds of Illinois; occasional elsewhere.

MEDICINE: Bark tea used by the Meskwaki to treat colds, smallpox, and internal pains.
REFERENCES: H. Smith 1928:235.

POLEMONIACEAE
Phlox Family

Phlox pilosa L.
Downy phlox
Native herb; occasional throughout Illinois in dry woods and prairies.

MEDICINE: Herb infusion used by the Meskwaki as a wash for eczema and to purify the blood.
REFERENCES: H. Smith 1928:235.

Polemonium reptans L.
Jacob's-ladder
Native herb; common throughout Illinois in moist or dry woods.

MEDICINE: Herb or roots used by the Meskwaki in combination with mayapple for urinary troubles or as a physic.
REFERENCES: H. Smith 1928:235.

POLYGALACEAE
Milkwort Family

Polygala polygama Walt.
Purple milkwort
Native herb; occasional on sandy waste ground in open woods in the northern one-half of Illinois.

MEDICINE: Herb boiled by the Montagnais to make cough medicine.
REFERENCES: Vogel 1970:372.

Polygala senega L.
Seneca snakeroot
Native herb; occasional in prairies and dry woods in the northern three-fourths of Illinois.

MEDICINE: Root tea used by the Meskwaki as a remedy for heart trouble and as a tonic by the Ojibwa; root used by the Seneca for snakebite; boiled plant used by the Nishinams for diarrhea. Recent--rootstock used as a diaphoretic, diuretic, expectorant, and emetic.
CEREMONY: Used by the Ojibwa as a good luck charm when traveling.
REFERENCES: Densmore 1928:291; Lewis and Elvin-Lewis 1977:287, 347; H. Smith 1928:236; Vogel 1970:371-73; Tehon 1951:89.

POLYGONACEAE
Knotweed Family

Polygonum aviculare L.
Knotweed
Introduced herb; common throughout Illinois on waste ground.

MEDICINE: Recent-- herb has been used as an astringent, seeds as aromatic, purgative, and emetic.
REFERENCES: Tehon 1951:90.

Polygonum careyi Olney
Knotweed
Native herb; occurs on sandy soil; rare.

MEDICINE: Herb tea used by the Potawatomi to treat colds.
REFERENCES: H. Smith 1933:72.

Polygonum coccineum Muhl.
Water smartweed
Native herb; occasional throughout Illinois on wet ground.

SYNONYMS: *P. muhlenbergii* (Meisn.)S.Wats.
FOOD: Young shoots eaten in spring by the Sioux.
MEDICINE: Herb tea used by the Ojibwa for stomach pain and by the Meskwaki for mouth sores.
SMOKING: Dried flowers smoked by the Ojibwa to attract deer.
REFERENCES: H. Smith 1928:236, 1932:381,431; Yanovsky 1936:20.

Polygonum erectum L.
Knotweed
Native herb; occasional to common throughout Illinois on waste ground.

FOOD: Seeds found in numerous Illinois archaeological sites; no recorded Historic Indian use.
REFERENCES: Asch and Asch 1981.

Polygonum fluitans Eaton
Water smartweed
Native herb; occasional in shallow water in the northern one-half of Illinois; rare elsewhere.

MEDICINE: Root used by the Potawatomi for "medicine."
REFERENCES: H. Smith 1933:72.

Polygonum hydropiper L.
Water pepper
Introduced herb; occasional throughout Illinois on wet ground.

MEDICINE: Recent--the herb has been used externally as a counterirritant, internally as a stimulant and diaphoretic.
DYE: Recent--the herb can be used for yellow and gold.
REFERENCES: Brooklyn Botanic Garden 1964:21; Lust 1974:561; Tehon 1951:90.

Polygonum lapathifolium L.
Pale smartweed
Native herb; common throughout Illinois on wet ground.

MEDICINE: Herb tea used by the Potawatomi to cure fevers.
REFERENCES: H. Smith 1933:72.

Polygonum pensylvanicum L.
Common smartweed
Native herb; occasional throughout Illinois on wet ground.

MEDICINE: Herb tea used by the Meskwaki to cure diarrhea, by the Menominee to stop bleeding of the mouth and to heal the effects of childbirth, and by the Ojibwa to treat epilepsy.
REFERENCES: Gilmore 1933:129; H. Smith 1923:47, 1928:236.

Polygonum persicaria L.
Lady's thumb
Introduced herb; common throughout Illinois on waste ground.

MEDICINE: Used by the Ojibwa to treat stomach pains.
REFERENCES: Densmore 1928:291.

Polygonum punctatum Ell.
Smartweed
Native herb; common throughout Illinois on wet ground.

MEDICINE: Used by the Ojibwa for "medicine." Recent--herb used externally as a counterirritant, internally as a diaphoretic and stimulant. The leaves contain 7% calcium oxalate and are poisonous.
REFERENCES: Densmore 1928:291; Lewis and Elvin-Lewis 1977:34; Tehon 1951:90.

Rumex acetosella L.
Sour dock
Introduced herb; common throughout Illinois on waste ground.

FOOD: Plant eaten by the Iroquois. Recent--the plants are toxic to livestock in large quantities.
MEDICINE: Recent--powdered root once used as an abrasive dentifrice; used in Europe as a laxative, tonic, and alterative.
REFERENCES: Lewis and Elvin-Lewis 1977:34,284,347; Waugh 1916:118; Yanovsky 1936:20.

Rumex crispus L.
Curly dock
Introduced herb; common throughout Illinois on waste ground.

MEDICINE: Root used by the Ojibwa for healing cuts and skin eruptions. Recent--roots used as a tonic, astringent, and laxative.
DYE: Recent--roots can be used for black.
REFERENCES: Brooklyn Botanic Garden 1964:65; Densmore 1928:292; H. Smith 1932:381; Tehon 1951:98.

Rumex obtusifolius L.
Bitter dock
Introduced herb; occasional throughout Illinois on waste ground.

MEDICINE: Root tea used by the Ojibwa for cuts and ulcers.
DYE: Recent--roots can be used for dark yellow.
REFERENCES: Brooklyn Botanic Garden 1964:22; Densmore 1928:292; Lust 1974:553.

Rumex orbiculatus Gray
Water dock
Native herb; occasional in moist soil in the northern two-thirds of Illinois.

SYNONYMS: *R. britannica* L. (as used by H. Smith 1928:237).
MEDICINE: Root tea used by the Meskwaki as an antidote for poisoning.
REFERENCES: H. Smith 1928:237.

PORTULACACEAE
Purslane Family

Claytonia virginica L.
Spring beauty
Native herb; common throughout Illinois in woods.

FOOD: Roots eaten by eastern Indians.
REFERENCES: Waugh 1916:120; Yanovsky 1936:24.

PRIMULACEAE
Primrose Family

Anagallis arvensis L.
Scarlet pimpernel
Introduced herb; occurs on waste ground throughout Illinois; uncommon.

MEDICINE: Recent--leaves and flowers have been used as an expectorant and nervine in the treatment of rheumatism; plant juices used as a gargle for toothache. The leaves are toxic to livestock.
REFERENCES: Lewis and Elvin-Lewis 1977:35,252; Tehon 1951:19.

Trientalis borealis (Raf.)Pursh
Star-flower
Native herb; occurs in moist woods in northeastern Illinois; rare.

SYNONYMS: *T. americana* Pursh
MEDICINE: Herb tea drunk by Quebec Indians for sickness and tuberculosis.
CEREMONY: Root included by Ojibwa in hunting medicine to attract deer.
REFERENCES: H. Smith 1932:431; Vogel 1970:271.

RANUNCULACEAE
Buttercup Family

Actaea pachypoda Ell.
White baneberry
Native herb; occasional throughout Illinois in rich woods.

SYNONYMS: *A. alba* (L.)Mill.
MEDICINE: Root infusion used by the Meskwaki to relieve pain during childbirth and for genitourinary problems and by the Ojibwa for convulsions. All parts of the plant are toxic.
REFERENCES: Gilmore 1919:130; Lewis and Elvin-Lewis 1977:30; H. Smith 1928:237.

Actaea rubra (Ait.)Willd.
Red baneberry
Native herb; occurs in rich woods in northern Illinois; uncommon.

MEDICINE: Root infusion used by the Potawatomi and Ojibwa to purge afterbirth and roots eaten for stomach troubles; used by the Meskwaki for lung trouble and fever. All parts of the plant are toxic.
REFERENCES: Densmore 1928:286; Lewis and Elvin-Lewis 1977:30; H. Smith 1928:195, 1932:382, 1933:74; Vogel 1970:235-36.

Anemone canadensis L.
Meadow anemone
Native herb; occasional in open woods and moist prairies in the northern two-thirds of Illinois.

MEDICINE: Root eaten by the Ojibwa to clear the throat, leaves used as a styptic for hemorrhage, root infusion used as a mouthwash and for sores; used by the Meskwaki for eye problems and used by the Indians of the Missouri River region for "medicine."
REFERENCES: Gilmore 1919:30, 1933:130; H. Smith 1928:238, 1932:382; Vogel 1970:271-72.

Anemone cylindrica Gray
Thimbleweed
Native herb; occasional in open woods and prairies in the northern two-thirds of Illinois.

MEDICINE: Root infusion used by the Ojibwa for lung congestion and tuberculosis, seeds used by the Meskwaki as a poultice for burns; root tea used for dizziness, headaches, and insanity.
CEREMONY: Used by the Indians of the Missouri River region as a good luck charm.
REFERENCES: Gilmore 1919:30; H. Smith 1928:238, 1932:383; Vogel 1970:271-72.

Anemone patens L.
Pasque flower
Native herb; occurs in prairies in northern Illinois; uncommon.

SYNONYMS: *Anemone ludoviciana* Nutt., *Pulsatilla hirsutissima* (Pursh)Britt.
MEDICINE: Used for headache by the Ojibwa. Recent--considered useful as an irritant, diuretic, expectorant,and emmenagogue.
REFERENCES: Tehon 1951:19; Vogel 1970:272.

Anemone virginiana L.
Tall anemone
Native herb; common throughout Illinois in open, usually dry, woods.

MEDICINE: Root used for poultices by the Menominee; seedpods burned by the Meskwaki to revive an unconscious person.
REFERENCES: H. Smith 1923:48, 1928:238; Vogel 1970:271.

Anemonella thalictroides (L.)Spach
Rue anemone
Native herb; occasional throughout Illinois in dry or moist open woods.

MEDICINE: Used for "medicine" by the Cherokee Indians.
REFERENCES: Vogel 1970:102.

Aquilegia canadensis L.
Columbine
Native herb; occasional to common throughout Illinois in rocky woods.

MEDICINE: Root used by the Ojibwa and Meskwaki for stomach troubles. The plant and seeds are poisonous.
CEREMONY: Used as a love charm by the Indians of the Missouri River region.
SMOKING: Seeds used to perfume a mixture smoked by the Menominee and Ojibwa.
REFERENCES: Epstein 1981:96; Gilmore 1919:30; Lewis and Elvin-Lewis 1977:30; H. Smith 1928:238, 1932:383.

Caltha palustris L.
Marsh marigold
Native herb; occasional in wet meadows in the northern two-thirds of Illinois.

FOOD: Leaves and stems boiled as potherbs by the Ojibwa, Iroquois, and Menominee.
MEDICINE: Root used to treat sores by the Ojibwa.
REFERENCES: Densmore 1928:288; Gilmore 1933:130; H. Smith 1923:70, 1932:408; Waugh 1916:117; Yanovsky 1936:25.

Cimicifuga racemosa (L.)Nutt.
Black cohosh
Native herb; occurs locally in woods throughout Illinois.

MEDICINE: Used by the Winnebagos to treat rheumatism. More recently, dried roots used as a sedative, emmenagogue, and alterative; demonstrates experimental hypoglycemic effect.
REFERENCES: Lewis and Elvin-Lewis 1977:219; Tehon 1951:40; Vogel 1970:59,133,291,368,370.

Clematis viorna L.
Leatherflower
Native vine; occurs in moist soil, particularly at the edge of woods, in southern Illinois; rare.

MEDICINE: Root tea used by the Meskwaki for any type of illness. (This is probably an incorrect identification since only *C. virginiana* L. and *C. verticillaris* DC. occur in Wisconsin.)
REFERENCES: H. Smith 1928:239.

Clematis virginiana L.
Virgin's bower
Native vine, occasional to common throughout Illinois on moist soil.
MEDICINE: Plant is poisonous and a violent purgative.
DYE: Recent--twigs and leaves can be used for yellow.
REFERENCES: Lust 1974:563.

Delphinium ajacis L.
Rocket larkspur
Introduced herb; cultivated and occasionally escaped.

MEDICINE: Plant and seeds are poisonous to humans and livestock. Seeds have been used in and against body parasites.
DYE: Recent--flowers can be used for green.
REFERENCES: Lewis and Elvin-Lewis 1977:30-31; Lust 1974:557; Tehon 1951:47.

Hepatica nobilis Schreb. var. *acuta* (Pursh)Steyerm.
Liverleaf
Native herb; occasional throughout Illinois in rich woods.

SYNONYMS: *H. acutiloba* DC.
MEDICINE: Root tea used by the Menominee for female problems and by the Meskwaki for crossed eyes.
REFERENCES: H. Smith 1923:49, 1928:239; Tehon 1951:64; Vogel 1970:317.

Hepatica nobilis Schreb. var. *obtusa* (Pursh)Steyerm.
Liverleaf
Native herb; occurs in rich woods in northeastern Illinois; uncommon.

SYNONYMS: *H. americana* (DC.)Ker, *H. triloba* Beck, *H. triloba* L.
MEDICINE: Root used by the Ojibwa for convulsions; root and leaf infusion used by the Potawatomi for vertigo.
DYE: Used by the Potawatomi for dyeing mats and baskets.
CEREMONY: Used as a charm by the Ojibwa.
REFERENCES: Densmore 1928:289; H. Smith 1933:75; Tehon 1951:64; Vogel 1970:317.

Hydrastis canadensis L.
Goldenseal
Native herb; occasional throughout Illinois in rich woods.

MEDICINE: Used as an inhalant for catarrh by the Meskwaki and by Indians of eastern North America and early settlers for liver and stomach ailments. Recent--the leaves and roots were used as a tonic, especially in catarrhal affectations; the dried rhizome astringent and source of an antibiotic.
REFERENCES: Lewis and Elvin-Lewis 1977:276,343,360; H. Smith 1928:194; Tehon 1951:66; Vogel 1970:311-12.

Ranunculus abortivus L.
Small-flowered crowfoot

Native herb; common throughout Illinois in fields and moist ground.
MEDICINE: Root used as a styptic by the Meskwaki.
REFERENCES: H. Smith 1928:239.

Ranunculus acris L.
Tall buttercup
Introduced herb; occasional in fields and roadsides in the northern one-half of Illinois.

DYE: Recent--tops can be used for purple.
REFERENCES: Lust 1974:562.

Ranunculus flabellaris Raf.
Yellow water-crowfoot
Native herb; occasional throughout Illinois in swamps and ponds.

SYNONYMS: *R. delphinifolius* Torr.
MEDICINE: Flowers used by the Meskwaki to cause sneezing.
REFERENCES: H. Smith 1928:239-40; Vogel 1970:297.

Ranunculus pennsylvanicus L.f.
Bristly crowfoot
Native herb; occasional throughout Illinois on wet ground.

MEDICINE: Plant used as an astringent by the Potawatomi.
DYE: Plant used for yellow by the Ojibwa and Potawatomi and for red by the Ojibwa.
SMOKING: Seeds included by Ojibwa in hunting medicine smoked to attract deer.
REFERENCES: H. Smith 1932:383,426,431, 1933:75,123; Vogel 1970:297.

Ranunculus recurvatus Poir.
Crowfoot
Native herb; occasional throughout Illinois in wet woods.

DYE: Root used for red by the Menominee.
REFERENCES: H. Smith 1923:79.

Thalictrum dasycarpum Fisch. & Call.
Purple meadow-rue
Native herb; occasional throughout Illinois in moist woods and ravines.

MEDICINE: Roots used to treat fevers by the Ojibwa; leaves and seeds used by the Potawatomi for cramps and seeds used to make poultices more effective.
CEREMONY: Seeds used as love and good luck charms by the Potawatomi, as a love charm by the Indians of the Missouri River region.
SMOKING: Seeds smoked by Potawatomi while hunting to bring good luck or mixed with tobacco.
REFERENCES: Epstein 1981:135-36; Gilmore 1919:28; H. Smith 1932:383, 1933:75,123.

RHAMNACEAE
Buckthorn Family

Ceanothus americanus L.
New Jersey tea
Native shrub; occasional throughout Illinois in woods and thickets.

FOOD: Leaves used to make tea by Alabama Indians and by colonists during the Revolutionary war.
MEDICINE: Root tea used by Menominee for stomach troubles; by the Meskwaki for snakebite, as an emetic, and for eczema; and by the Ojibwa for constipation and for pulmonary troubles. Recent--the leaves and roots were used as an astringent and styptic.
REFERENCES: Gilmore 1919:102, 1933:136; Lewis and Elvin-Lewis 1977:391; H. Smith 1923:49,70, 1928:193,240-41,263; Tehon 1951:35; Vogel 1970:169; Yanovsky 1936:42.

Ceanothus ovatus Desf.
Inland Jersey tea
Native shrub; occurs on low dunes and sandy soil in northern Illinois; rare.

MEDICINE: Root used by the Ojibwa for coughs and as an emetic.
REFERENCES: Densmore 1928:288,340.

Rhamnus alnifolia L'her.
Alder buckthorn
Native shrub; occurs in bogs and wooded swamps in the northern one-half of Illinois; rare.

MEDICINE: Bark used by the Potawatomi as a physic and by the Meskwaki to treat constipation.
REFERENCES: H. Smith 1928:241, 1933:75.

Rhamnus frangula L.
Glossy buckthorn
Introduced shrub; occasional in woods and bogs in the northern one-half of Illinois.

MEDICINE: Recent--the bark has been used for constipation and "cleaning the blood." The fruits are poisonous.
DYE: Recent--bark can be used for bronze-brown.
REFERENCES: Lewis and Elvin-Lewis 1977:40,284; Brooklyn Botanic Garden 1964:68; Lust 1974:550.

ROSACEAE
Rose Family

Agrimonia gryposepala Wallr.
Tall agrimony
Native herb; occasional in woods and thickets in the northern two-thirds of Illinois.

MEDICINE: Root used by the Ojibwa for urinary problems and by the Meskwaki and Potawatomi to stop nosebleeds. Recent--used as an astringent.
REFERENCES: H. Smith 1928:241, 1932:383-84, 1933:76; Tehon 1951:16.

Amelanchier arborea (Michx.f.)Fern.
Shadbush
Native tree; occasional throughout Illinois on wooded hillsides and streambanks.

SYNONYMS: *A. canadensis* (L.)Medic.
FOOD: Berries eaten by the Ojibwa and Iroquois fresh, dried and boiled with meat, or made into pemmican.
REFERENCES: Densmore 1928:307; Waugh 1916:128; Yanovsky 1936:30.

Amelanchier humilis Wieg.
Juneberry
Native shrub; occasional in rocky or sandy soil in the northern one-fourth of Illinois.

SYNONYMS: *A. spicata* (Lam.)K.Koch
FOOD: Fruit eaten by the Iroquois and Potawatomi.
MEDICINE: Bark used by the Potawatomi as a tonic and by the Ojibwa for "medicine" and as a tea for expectant mothers.
REFERENCES: H. Smith 1933:76,107; Yanovsky 1936:30.

Amelanchier laevis Wieg.
Shadbush
Native tree; occasional on wooded slopes in the northern one-fourth of Illinois.

FOOD: Fruit eaten by the Ojibwa either fresh or dried.
MEDICINE: Bark used by the Ojibwa for "medicine."
REFERENCES: H. Smith 1932:384,408; Yanovsky 1936:30.

Amelanchier spp.
Shadbush
Native shrub; occasional throughout Illinois in woods.

TECHNOLOGY: Stems used by the Ojibwa to make arrows.
REFERENCES: Densmore 1928:147.

Aronia melanocarpa (Michx.)Ell.
Black chokeberry
Native shrub; occasional in bogs and moist woods in the northeastern one-fourth of Illinois.

SYNONYMS: *Pyrus melanocarpa* (Michx.)Willd.
FOOD: Berries eaten by the Potawatomi.
MEDICINE: Berries used by the Potawatomi to make a beverage to treat colds.
REFERENCES: H. Smith 1933:76,107.

Crataegus calpodendron (Ehrh.)Medic.
Hawthorn
Native tree; occasional throughout Illinois in woods and thickets.

SYNONYMS: *C. tomentosa* Duroi
FOOD: Fruit eaten by the Meskwaki either raw or cooked.
MEDICINE: Unripe fruit used by the Meskwaki to treat bladder troubles and by the Potawatomi for stomach problems.
REFERENCES: Lewis and Elvin-Lewis 1977:193,387,395; H. Smith 1928:241,263; Yanovsky 1936:31.

Crataegus pedicellata Sarg.
Hawthorn
Native tree; occasional throughout Illinois in thickets and along the borders of woods.

SYNONYMS: *C. coccinea* L.
FOOD: Fruit eaten by Indians in the eastern states either fresh or pressed into cakes with other fruits for winter use.
REFERENCES: Yanovsky 1936:30.

Crataegus mollis (T.&G.)Scheele
Red haw
Native tree; occasional throughout Illinois in woods.

FOOD: Fruit eaten by Indians in the eastern states.
REFERENCES: Yanovsky 1936:30.

Crataegus pruinosa (Wendl.)K.Koch
Hawthorn
Native tree; occasional throughout Illinois in thickets and rocky woods.

FOOD: Fruit eaten by the Iroquois.
REFERENCES: Waugh 1916:128; Yanovsky 1936:31.

Crataegus spp.
Hawthorn
Native tree; common throughout Illinois in many different habitats.

FOOD: Fruit eaten by many aboriginal groups. Recent--seeds used as a coffee substitute.
MEDICINE: Bark used by the Ojibwa as a vermifuge and for female disorders; root tea used for tuberculosis. Recent--many species are hypotonic and have an antiarrhythmic effect.
TECHNOLOGY: Thorns used by the Ojibwa as needles.
SMOKING: Bark included by the Ojibwa in a hunting medicine smoked to attract deer.
REFERENCES: Densmore 1928:289; Gilmore 1933:132; Lewis and Elvin-Lewis 1977:193,387,395; H. Smith 1932:384,422,431.

Filipendula rubra (Hill)Rob.
Queen-of-the-prairie
Native herb; occasional in fens in the northern one-half of Illinois.

MEDICINE: Root used by the Meskwaki to treat heart troubles.
CEREMONY: Used by the Meskwaki in a love medicine.
REFERENCES: H. Smith 1928:241.

Fragaria americana (Porter)Britt.
Hillside strawberry
Native herb; occasional to rare on wooded slopes in the northern one-half of Illinois.

SYNONYMS: *F. vesca* Porter
FOOD: Fruit eaten by many tribes either fresh, dried, or made into preserves; tea made from the leaves.
MEDICINE: Root used by the Potawatomi to treat stomachache. Leaf tea used by western Indians for diarrhea.
REFERENCES: Gilmore 1919:84; Lewis and Elvin-Lewis 1977:287; H. Smith 1933:76,107; Yanovsky 1936:30.

Fragaria virginiana Duchesne
Wild strawberry
Native herb; occasional to common throughout Illinois in open woods, prairies, and fields.

FOOD: Berries eaten by many tribes either fresh or made into jam for winter use; tea made from the leaves.

MEDICINE: Used by the Ojibwa for stomach trouble and to treat cholera. Recent--fruit is said to have astringent and diuretic properties.
REFERENCES: Densmore 1928:289,321; Gilmore 1919:84; H. Smith 1923:49,71, 1928:263, 1932:384; Tehon 1951:56; Yanovsky 1936:30.

Geum canadense Jacq.
White avens
Native herb; common throughout Illinois in woodlands.

MEDICINE: Root used by the Meskwaki for eczema and by the Ojibwa for female troubles.
REFERENCES: Densmore 1928:289; H. Smith 1928:194.

Geum rivale L.
Purple avens
Native herb; occurs in wet meadows in northeastern Illinois; rare.

FOOD: Decoction of the fragrant rootstock used as a beverage in the northern states and Canada.
MEDICINE: Recent--the root has been used as an astringent.
REFERENCES: Vogel 1970:388; Yanovsky 1936:31.

Geum strictum Ait.
Yellow avens
Native herb; occasional in bogs and moist thickets in the northern one-half of Illinois.

MEDICINE: Root tea used by the Ojibwa for soreness in the chest and coughs.
REFERENCES: Vogel 1970:388.

Geum triflorum Pursh
Prairie avens
Native herb; common in dry prairies in the northern one-sixth of Illinois.

SYNONYMS: *Sieversia ciliata* (Pursh)Rydb.
FOOD: Roots boiled by the Indians of British Columbia to make a beverage.
MEDICINE: Root used by the Ojibwa as a tonic and stimulant.
REFERENCES: Densmore 1928:293,362; Yanovsky 1936:32.

Gillenia stipulata (Muhl.)Baill.
Indian physic
Native herb; occasional in woods in the southern one-half of Illinois.

MEDICINE: Recent--the roots have been used as an emetic and stomach tonic.
REFERENCES: Lewis and Elvin-Lewis 1977:278; Tehon 1951:60.

Malus coronaria (L.)Mill.
Wild sweet crabapple
Native tree; occasional in woods in the southern one-half of Illinois.

FOOD: Fruit eaten by the Indians of the northern and eastern states. Seeds of the genus *Malus* are poisonous in large quantities.
MEDICINE: Recent--the bark and root are the source of an antibiotic.
REFERENCES: Lewis and Elvin-Lewis 1977:40,362; Parker 1910:94; Waugh 1916:129; Yanovsky 1936:1936:32.

Malus ioensis (Wood)Britt.
Iowa crabapple
Native tree; occasional to common throughout Illinois in woods and thickets.

SYNONYMS: *Pyrus ioensis* (Wood)Bailey
FOOD: Fruit eaten by the Meskwaki raw, dried, or made into jelly.
MEDICINE: Used by the Meskwaki to treat smallpox.
REFERENCES: H. Smith 1928:242,263; Yanovsky 1936:32.

Malus pumila Mill.
Apple
Introduced tree; cultivated and sometimes escaped.

SYNONYMS: *Pyrus malus* L.
FOOD: Often cultivated for the large edible fruit.
DYE: Recent--bark can be used for yellow.
REFERENCES: Brooklyn Botanic Garden 1964:68; Lust 1974:550.

Physocarpus opulifolius (L.)Maxim.
Ninebark
Native shrub; occasional on rocky slopes and banks and in moist swales in the northern one-half of Illinois.

MEDICINE: Bark tea used by the Menominee for female troubles and by the Ojibwa as an emetic in case of accidentally swallowing a waterstrider (insect).
REFERENCES: Gilmore 1933:132; H. Smith 1923:49.

Potentilla anserina L.
Silverweed
Native herb; occurs locally in northeastern Illinois on sandy beaches, gravel bars, and in interdunal ponds and meadows.

FOOD: Roots eaten by Indians in the Northwest.
REFERENCES: Yanovsky 1936:32.

Potentilla arguta Pursh
Prairie cinquefoil
Native herb; occasional in prairies in the northern one-half of Illinois.

SYNONYMS: *Drymocallis arguta* (Pursh)Ryb.
MEDICINE: Root used by the Ojibwa for headache and convulsions.
REFERENCES: Densmore 1928:289,338; Vogel 1970:225.

Potentilla norvegica L.
Rough cinquefoil
Introduced herb; occasional throughout Illinois on disturbed soil.

SYNONYMS: *P. monspeliensis* L.
MEDICINE: Plant used by the Ojibwa as a physic and for sore throats and by the Potawatomi for "medicine."
REFERENCES: Densmore 1928:291; H. Smith 1932:384, 1933:77.

Potentilla palustris (L.)Scop.
Marsh cinquefoil
Native herb; occurs in bogs in Cook, Lake, and McHenry counties.

MEDICINE: Root used by the Ojibwa for stomach cramps and dysentery.
REFERENCES: Densmore 1928:291; H. Smith 1932:385.

Prunus americana Marsh.
Wild plum
Native tree; occasional throughout Illinois in thickets and woodlands.

FOOD: Fruit used by Indians throughout its range either fresh, dried, or cooked.
MEDICINE: Root used by the Ojibwa as a vermifuge and disinfectant, by the Meskwaki to cure canker sores, and by the Indians of the Missouri River region to treat skin abrasions.
REFERENCES: Densmore 1928:291,321; Gilmore 1919:35,87; H. Smith 1928:242, 1932:128; Vogel 1970:176; Yanovsky 1936:32.

Prunus angustifolia Marsh.
Chickasaw plum
Native tree; occasional in thickets in the southern one-half of Illinois; uncommon elsewhere.

FOOD: Fruit eaten by Indians in the southern states.
REFERENCES: Yanovsky 1936:32.

Prunus hortulana Bailey
Wild goose plum
Native tree; occasional in thickets in the southern two-thirds of Illinois.

FOOD: Many similar species were used for food by Historic Indian groups.
REFERENCES: Yanovsky 1936:32-33.

Prunus munsoniana Wight & Hedrick
Wild goose plum
Native tree; occasional in thickets in the southern one-third of Illinois.

FOOD: Many similar species were used for food by Historic Indian groups.
REFERENCES: Yanovsky 1936:32-33.

Prunus nigra Ait.
Canada plum
Native tree; occasional along rivers and streams in the northern one-half of Illinois.

FOOD: Fruit eaten by Indians throughout its range either fresh or made into jam.
MEDICINE: Bark tea used by the Meskwaki to treat stomach troubles.
DYE: Used by the Ojibwa as a fixative in dyeing.
REFERENCES: H. Smith 1928:242,263, 1932:409,426; Vogel 1970:33; Waugh 1916:128; Yanovsky 1936:33.

Prunus pensylvanica L.f.
Pin cherry
Native tree; occasional on sandy soil in the northern one-third of Illinois.

FOOD: Fruit eaten by the Potawatomi, Ojibwa, Iroquois, and probably others.
MEDICINE: Bark tea used by the Potawatomi and Ojibwa for coughs and colds.
REFERENCES: H. Smith 1932:385, 1933:77; Waugh 1916:128; Yanovsky 1936:33.

Prunus persica (L.)Batsch
Peach
Introduced tree; cultivated and occasionally escaped.

FOOD: Because of the large, juicy, fruit, the peach was widely distributed by North American Indians shortly after its introduction by the Spanish.
DYE: Recent--the leaves can be used for yellow.
REFERENCES: Lust 1974:559; Sheldon 1980.

Prunus serotina Ehrh.
Wild black cherry
Native tree; common throughout Illinois in woods and fencerows.

FOOD: The fruit was eaten by Indians throughout its range either fresh or dried; an infusion of twigs used as a beverage.
MEDICINE: Bark tea used by the Ojibwa for coughs, colds, and indigestion, by the Menominee for "medicine," and by the Potawatomi as a seasoner of other medicines. Recent--the bark has been used as a tonic, astringent and sedative. This is the most dangerous of the species of wild cherry native to eastern North America; all parts, especially the leaves and seeds, contain compounds which can result in cyanide poisoning.
DYE: Recent--young leaves and green fruits used for yellow and orange.
REFERENCES: Brooklyn Botanic Garden 1973:19; Densmore 1928:291,317; Lewis and Elvin-Lewis 1977:40; H. Smith 1923:49,71, 1932: 385,409, 1933:108; Yanovsky 1936:33.

Prunus susquehanae Willd.
Sand cherry
Native tree; local on sandy soil in the northern one-sixth of Illinois.

SYNONYMS: *P. pumila* L.
FOOD: Fruit eaten by Indians in the northern states either fresh or preserved.
REFERENCES: H. Smith 1923:71, 1932:409; Yanovsky 1936:33.

Prunus virginiana L.
Common chokecherry
Native tree; occasional in woods and thickets in the northern one-half of Illinois; rare elsewhere.

SYNONYMS: *Padus nana* (Duroi)Roemer
FOOD: Fruit eaten by Indians throughout its range fresh, dried, or made into pemmican; bark and twigs used to make a beverage.
MEDICINE: Inner bark pounded by the Menominee for a poultice and bark tea used for diarrhea; used by the Meskwaki as a sedative and for stomach troubles; by the Ojibwa for lung troubles; and by the Potawatomi for eyewash.
REFERENCES: Densmore 1928:317,321; Gilmore 1919:88; H. Smith 1923:50,71, 1928:242,263, 1932:385,408, 1933:77,108.

Rosa blanda Ait.
Meadow rose
Native shrub; occasional in thickets, woods, and open areas in the northern one-third of Illinois; rare elsewhere.

MEDICINE: Rose hip skin used by the Meskwaki and Ojibwa for stomach trouble and hemorrhoids; powdered flowers used by the Ojibwa for heartburn; root tea used by the Potawatomi for headaches and lumbago.
REFERENCES: H. Smith 1928:242, 1932:385, 1933:78.

Rosa carolina L.
Pasture rose
Native shrub; common throughout Illinois in prairies, fields, and dry woods.

SYNONYMS: *R. humilis* Marsh.
MEDICINE: Rose hip skin used by the Menominee to treat stomach troubles.
REFERENCES: H. Smith 1923:50.

Rosa setigera Michx.
Prairie rose
Native shrub; occasional to common throughout Illinois in woods, thickets, and clearings.

SMOKING: Inner bark smoked by the Indians of the Missouri River region.
REFERENCES: Epstein 1981:100; Gilmore 1919:33.

Rosa suffulta Greene
Rose
Native shrub; occasional in thickets and woods in the northern two-thirds of Illinois.

SYNONYMS: *R. arkansana* Porter var. *suffulta* (Greene)Cockerell
MEDICINE: Root used by the Ojibwa in a tonic.
REFERENCES: Densmore 1928:364.

Rosa spp.
Wild rose
Native shrub; common throughout Illinois in many habitats.

FOOD: No species occurring in Illinois is recorded as having been used for food by Indians of eastern North America. However, the fruits of numerous other species were eaten in other regions.
REFERENCES: Yanovsky 1936:33–34.

Rubus allegheniensis Porter
Common blackberry
Native shrub; common throughout Illinois on roadsides and in thickets and woods.

FOOD: Fruit was eaten by the Ojibwa, Menominee, Meskwaki, Potawatomi, and others, either fresh or dried.
MEDICINE: Root tea used by the Ojibwa to treat diarrhea, by the Potawatomi and Meskwaki for sore eyes, and by the Meskwaki for stomach troubles. Recent-- the fruit, root, and bark were used as a mild astringent and in fruit syrup.
REFERENCES: Gilmore 1933:133; H. Smith 1923:50,71, 1928:243,264, 1932:409, 1933:79,108; Tehon 1951:98; Vogel 1970:282; Yanovsky 1936:33.

Rubus alumnus Bailey
Blackberry
Native shrub; occurs along railroad tracks in Jackson and Wabash counties; rare.

FOOD: Many similar species were used for food by Historic Indian groups.
REFERENCES: Yanovsky 1936:34-35.

Rubus argutus Link
High-bush blackberry
Native shrub; common throughout Illinois in woods and thickets.

SYNONYMS: *R. ostryifolius* Rydb., *R. blakei* Bailey
FOOD: Many similar species were used for food by Historic Indian groups.
REFERENCES: Yanovsky 1936:34-35.

Rubus avipes Bailey
Blackberry
Native shrub; occurs in thickets in northern Illinois; rare.

FOOD: Many similar species were used for food by Historic Indian groups.
REFERENCES: Yanovsky 1936:34-35.

Rubus enslenii Tratt.
Arching dewberry
Native shrub; occurs in rocky woods in Randolph County; rare.

FOOD: Many similar species used for food by Historic Indian groups.
REFERENCES: Yanovsky 1936:34-35.

Rubus flagellaris Willd.
Dewberry
Native shrub; occasional to common throughout Illinois in fields, edges of woods and along roadsides.

FOOD: Fruit eaten by the Indians of New York State.
REFERENCES: Yanovsky 1936:34.

Rubus frondosus Bigel.
Blackberry
Native shrub; occasional throughout Illinois in woods and thickets.

FOOD: Fruit eaten by the Ojibwa.
MEDICINE: Used by the Ojibwa for lung troubles.
REFERENCES: Densmore 1928:292,321; Yanovsky 1936:34.

Rubus hispidus L.
Swamp dewberry
Native shrub; local in meadows or low woods in the northern one-half of Illinois.

FOOD: Black seedy fruit sometimes eaten by eastern Indians.
REFERENCES: Yanovsky 1936:34.

Rubus occidentalis L.
Black raspberry
Native shrub; common throughout Illinois in thickets and woods and on bluffs and roadsides.

FOOD: Fruit eaten by Indians throughout its range, either fresh or dried; young shoots and sprouts eaten like rhubarb; leaves and bark of roots used for beverage.
MEDICINE: Root used by the Ojibwa to treat sore eyes and for female troubles, by the Ojibwa and Menominee for tuberculosis, and by the Indians of the Missouri River region to treat bowel trouble in children.
REFERENCES: Densmore 1928:292; Gilmore 1919:32,84, 1933:133; H. Smith 1923:50,71, 1928:264; Waugh 1916:127; Yanovsky 1936:34.

Rubus odoratus L.
Purple flowering raspberry
Native shrub; occurs in woods and thickets in northeastern Illinois; rare.

FOOD: Dry, insipid fruit eaten by the Iroquois.
REFERENCES: Parker 1910:96, Waugh 1916:127; Yanovsky 1936:34.

Rubus pennsylvanicus Poir.
Blackberry
Native shrub; occasional to common throughout Illinois in thickets and woods.

FOOD: Many similar species used for food by Historic Indian groups.
REFERENCES: Yanovsky 1936:34–35.

Rubus pubescens Raf.
Dwarf raspberry
Native shrub; local in bogs in northeastern Illinois.

SYNONYMS: *R. triflorus* Richards.
FOOD: Fruit eaten by the Iroquois.
REFERENCES: Waugh 1916:127; Yanovsky 1936:35.

Rubus schneideri Bailey
Bristly blackberry
Native shrub; occurs in sandy swales in Kankakee County; rare.

FOOD: Many similar species used for food by Historic Indian groups.
REFERENCES: Yanovsky 1936:34–35.

Rubus spp.
Blackberry
Native shrub; common throughout Illinois in numerous habitats.

DYE: Recent--young shoots used for light gray.
REFERENCES: Brooklyn Botanic Garden 1964:32; Lust 1974:552.

Rubus strigosus Michx.
Red raspberry
Native shrub; bogs and swampy woods in the northern one-fourth of Illinois.

SYNONYMS: *R. idaeus* L. var. *aculeatissimus* Robins. & Fern.

FOOD: Fruit eaten by Indians throughout its range, either fresh or dried; shoots peeled and eaten; leaves and twigs used to make a beverage.
MEDICINE: Root used by the Ojibwa for stomach pain, by the Ojibwa and Potawatomi for measles and eye problems, female troubles, and dysentery, and by the Menominee and Meskwaki as a seasoner of other medicines.
REFERENCES: Densmore 1928:291,317,321; Gilmore 1919:84, 1933:132; H. Smith 1923:50,71, 1932:386,410, 1933:79,109; Yanovsky 1936:35.

Rubus trivialis Michx.
Southern dewberry
Native shrub; occasional in fields and waste ground in the southern one-fourth of Illinois.

FOOD: Fruit eaten by Indians in the eastern states.
REFERENCES:Yanovsky 1936:35.

Sorbus aucuparia L.
European mountain ash
Introduced tree; occasional in bogs and swamps in the northern one-fifth of Illinois.

MEDICINE: Supposedly used for colds by the Potawatomi; possibly confused with the native *S. americana*.
DYE: Recent--bark used for gray; fruits for greenish-yellow.
REFERENCES: Brooklyn Botanic Garden 1973:36; H. Smith 1933:78.

Sorbus americana Marsh.
Mountain ash
Native tree; occurs in rocky woods in Ogle County; rare.

SYNONYMS: *Pyrus americana* (Marsh.)DC.
MEDICINE: Used by the Potawatomi and Ojibwa for "medicine."
TECHNOLOGY: Pliable wood used by the Ojibwa for canoe ribs, snowshoe frames, etc.
REFERENCES: Reagan 1928; H. Smith 1933:78.

Spiraea alba Duroi
Meadow-sweet
Native shrub; occasional on moist ground in the northern four-fifths of Illinois.

SYNONYMS: *S. salicifolia* Gray
MEDICINE: Used by the Meskwaki for diarrhea.
CEREMONY: Used as a trapping medicine by the Ojibwa.
SMOKING: Wood used by the Ojibwa for pipestems.
REFERENCES: H. Smith 1928:243, 1932:386; Vogel 1970:119.

Spiraea tomentosa L.
Hardhack
Native shrub; occurs in bogs and moist thickets in northeastern Illinois; rare.

MEDICINE: Used by the Ojibwa to make a medicinal beverage for any illness during pregnancy. Fruit decoction used by other Indians for diarrhea. Recent--bark and leaves considered useful as internal and topical astringent.
REFERENCES: H. Smith 1932:386; Vogel 1970:314.

RUBIACEAE
Madder Family

Cephalanthus occidentalis L.
Buttonbush
Native shrub; occasional throughout Illinois on wet ground.

MEDICINE: Inner bark used by the Meskwaki as an emetic. Recent--bark used in the treatment of fevers and chewed to relieve toothache.
REFERENCES: Lewis and Elvin-Lewis 1977:256; H. Smith 1928:243; Vogel 1970:247.

Galium aparine L.
Goosegrass
Native herb; common throughout Illinois in woods.

FOOD: Seeds used as a coffee substitute.
MEDICINE: Herb tea used by the Ojibwa for kidney and skin troubles; by the Meskwaki as an emetic. Recent--herb used as a diuretic and refrigerant.
REFERENCES: Gilmore 1933:141; Lewis and Elvin-Lewis 1977:192,387; H. Smith 1928:243, 1932:386; Vogel 1970:213.

Galium concinnum T.&G.
Shining bedstraw
Native herb; common throughout Illinois in dry woods.

MEDICINE: Herb tea used by the Meskwaki for bladder and kidney troubles.
REFERENCES: H. Smith 1928:244; Vogel 1970:213.

Galium tinctorium L.
Stiff bedstraw
Native herb; occasional in swamps and marshes in northeastern Illinois.

MEDICINE: Herb tea used by the Ojibwa for respiratory problems.
DYE: Recent--roots used for red.
REFERENCES: Lust 1974:554; H. Smith 1932:386; Vogel 1970:213.

Galium trifidum L.
Small bedstraw
Native herb; occurs on moist ground in northeastern Illinois; rare.

MEDICINE: Herb tea used by the Menominee for kidney troubles. Recent--herb used the same as *G. aparine* and in nerve disorders. The species contains an anticoagulant.
REFERENCES: Lewis and Elvin-Lewis 1977:192; H. Smith 1923:51; Tehon 1951:58; Vogel 1970:213.

Mitchella repens L.
Partridge-berry
Native herb; occasional in rocky or swampy woods in northern and southern Illinois; rare in the central counties.

FOOD: Berries eaten by Indians in the eastern states and Texas.
MEDICINE: Herb tea used by the Menominee to cure insomnia and by the Ojibwa as a "medicine." Recent--herb used as an astringent, diuretic, and parturifacient.

SMOKING: Smoked by the Ojibwa.
REFERENCES: Gilmore 1933:141; Reagan 1928:239; H. Smith 1923:51; Vogel 1970:344; Yanovsky 1936:57.

RUTACEAE
Rue Family

Ptelea trifoliata L.
Hop-tree
Native shrub; occasional throughout Illinois in open woods.

MEDICINE: Root bark used by the Menominee as a "sacred medicine" and as a seasoner of other medicines and by the Meskwaki for lung troubles. Recent--fruit, leaves, and root bark used as a tonic.
REFERENCES: H. Smith 1923:51, 1928:244; Tehon 1951:93; Vogel 1970:318.

Xanthoxylum americanum Mill.
Prickly ash
Native tree; occasional in woods in the northern half of Illinois; rare elsewhere.

SYNONYMS: *Zanthoxylum americanum* Mill.
MEDICINE: Root bark used by the Menominee in poultices and berries used for bronchial diseases, sores, and seasoning; bark tea used by the Ojibwa for sore throats; root used by the Potawatomi for gonorrhea. Recent--berries used as a tonic, stimulant, and diaphoretic.
REFERENCES: Densmore 1928:294; H. Smith 1923:51, 1932:387, 1933:80; Tehon 1951:120; Vogel 1970:352-54.

SALICACEAE
Willow Family

Populus balsamifera L.
Balsam poplar
Native tree; occurs on dunes near Lake Michigan; rare in Illinois.

SYNONYMS: *P. candicans* Ait.
MEDICINE: Resin used by the Potawatomi and Ojibwa as a salve for colds, congestion, and catarrh, by the Ojibwa for heart trouble and frostbite, and by the Menominee to treat head colds. Recent--leaf buds used as a tonic, stimulant, and expectorant. The genus *Populus* is a natural source of salicin and salicylic acid, disinfectant, antiinflammatory, antipyretic, antirheumatic, and analgesic.
REFERENCES: Densmore 1928:291; Gilmore 1933:126; Lewis and Elvin-Lewis 1977:151,308; H. Smith 1923:52, 1932:387, 1933:80-81; Vogel 1970:351-52.

Populus deltoides Marsh.
Cottonwood
Native tree; common throughout Illinois along streams and in low ground.

FOOD: Buds and seeds eaten by the Ojibwa.
MEDICINE: Down from buds used by the Ojibwa as an absorbent in open sores.
REFERENCES: Reagan 1928:243; Vogel 1970:77; Yanovsky 1936:16.

Populus grandidentata Michx.
Large-toothed aspen
Native tree; occasional to common in disturbed areas and riverbanks in the northern one-half of Illinois.

FOOD: Cambium eaten by the Ojibwa.
MEDICINE: Root used by the Ojibwa as a hemostatic.
REFERENCES: H. Smith 1932:388,410; Yanovsky 1936:16.

Populus tremuloides Michx.
Quaking aspen
Native tree; common in the low ground of woods, marshes, and bogs in northern Illinois.

FOOD: Cambium and sap eaten by the Ojibwa and in the Northwest.
MEDICINE: Bark used by the Ojibwa for poultices and female troubles; buds boiled in fat by the Meskwaki to make ointment for coughs and colds. Recent--used as a bitter tonic and feeble antiperiodic.
REFERENCES: Densmore 1928:291,320; H. Smith 1928:245, 1932:388; Vogel 1970:119,202; Yanovsky 1936:16.

Salix discolor Muhl.
Pussy willow
Native shrub or small tree; occasional throughout Illinois in marshes and swamps.

MEDICINE: A "universal remedy" of the Potawatomi; root tea used by the Ojibwa for hemorrhage and cambium used for indigestion; a stem ash decoction used by the Comanches for sore eyes. This genus is a natural source of salicin and salicylic acid (see *Populus balsamifera*).
REFERENCES: Densmore 1928:292; Lewis and Elvin-Lewis 1977:150,152; H. Smith 1933:81; Vogel 1970:393.

Salix fragilis L.
Crack willow
Introduced tree; planted and occasionally escaped.

MEDICINE: Bark used by the Ojibwa as a styptic and a poultice for sores.
REFERENCES: Densmore 1928:292; H. Smith 1932:388; Vogel 1970:393.

Salix humilis Marsh.
Prairie willow
Native shrub; occasional throughout Illinois in prairies.

SYNONYMS: *S. tristis* Ait.
MEDICINE: Root tea used by the Menominee for colic and diarrhea, by the Meskwaki for diarrhea and enemas, and by the Potawatomi as a universal remedy.
REFERENCES: Densmore 1928:292; H. Smith 1923:52, 1928:245, 1933:81; Vogel 1970:393.

Salix interior Rowlee
Sandbar willow
Native tree; common throughout Illinois on streambanks.

SYNONYMS: *S. longifolia* Muhl.
TECHNOLOGY: Branches used by the Ojibwa and Omaha for weaving baskets.
REFERENCES: Gilmore 1919:21, 1933:126.

Salix lucida Muhl.
Shining willow
Native tree; common in bogs and wet, sandy areas in the northern one-fourth of Illinois.

MEDICINE: Bark used by the Ojibwa as an external treatment for sores.
SMOKING: Bark included by the Ojibwa in a kinnikinnick mixture.
REFERENCES: Densmore 1928:292; H. Smith 1932:365,388.

Salix nigra Marsh.
Black willow
Native tree; common throughout Illinois in streams.

MEDICINE: Recent--bark used as an astringent and source of salicin.
TECHNOLOGY: Fibers used by the Ojibwa, Menominee, Ottawa, and Winnebago for making bags, pouches, fish nets, and cordage.
SMOKING: Inner bark added to smoking mixture by Plains Indians.
REFERENCES: Epstein 1981:103; Tehon 1951:101; Vogel 1970:393; Whitford 1941:11.

Salix pedicellaris Pursh
Bog willow
Native shrub; occurs in bogs in the northern one-third of Illinois; rare.

MEDICINE: Bark used by the Ojibwa to treat stomach trouble.
REFERENCES: Densmore 1928:292; H. Smith 1932:389.

Salix petiolaris Sm.
Petioled willow
Native shrub; occasional in low prairies, rich woods, marshes, and bogs in the northern one-fourth of Illinois.

SYNONYMS: *S. gracilis* Anderss. var. *textoris* Fern.
MEDICINE: Bark tea used by the Potawatomi and Ojibwa as a "universal remedy."
REFERENCES: Densmore 1928:292; H. Smith 1933:82.

SANTALACEAE
Sandalwood Family

Comandra richardsiana Fern.
False toadflax
Native herb; occasional throughout Illinois in woods and prairies.

SYNONYMS: *C. umbellata* (L.)Nutt.
MEDICINE: Leaf tea used by the Meskwaki for pain in the lungs.
REFERENCES: H. Smith 1928:246.

SAPOTACEAE
Sapodilla Family

Bumelia lanuginosa (Michx.)Pers.
Woolly buckthorn
Native tree; occurs in dry woods and thickets in Hardin, Pulaski, and Monroe counties; rare.

FOOD: Fruit eaten by Indians in the southern states.
MEDICINE: Ground bark source of mucilaginous substance used by New Mexico Indians as a chewing gum.
REFERENCES: Yanovsky 1936:52.

SARRACENIACEAE
Pitcher-plant Family

Sarracenia purpurea L.
Pitcher-plant
Native herb; occurs in bogs in northeastern Illinois; rare.

MEDICINE: Root or plant used for "medicine" by the Ojibwa, Potawatomi, and Menominee. Recent--root used as a bitter tonic and stomachic.
TECHNOLOGY: Leaves used by the Potawatomi in "old times" as drinking cups.
REFERENCES: H. Smith 1923:52, 1932:389, 1933:82,123; Tehon 1951:103; Vogel 1970:95.

SAXIFRAGACEAE
Saxifrage Family

Heuchera hirsuticaulis (Wheelock)Rydb.
Alum-root
Native herb; occasional in dry woods in the southern half of Illinois.

SYNONYMS: *H. americana* L.
MEDICINE: Powdered root used by some Indians of eastern North America on ulcers and wounds; leaves used by the Meskwaki as an astringent for healing sores. The plant, especially the root, contains a powerful astringent.
REFERENCES: Lewis and Elvin-Lewis 1977:352; H. Smith 1928:246.

Hydrangea arborescens L.
Hydrangea
Native shrub; common in woods in the southern four-fifths of Illinois; rare elsewhere.

MEDICINE: Recent--roots used as a diaphoretic and diuretic and for treating indigestion.
REFERENCES: Lewis and Elvin-Lewis 1977:46,273; Tehon 1951:65.

Mitella diphylla L.
Bishop's cap
Native herb; occasional in wooded ravines in the northern one-half of Illinois; rare elsewhere.

CEREMONY: Seeds used ceremonially by the Menominee.
REFERENCES: H. Smith 1923:81.

Ribes americanum Mill.
Wild black currant
Native shrub; common in moist woods in the northern one-third of Illinois.

SYNONYMS: *R. floridum* L'Her.
FOOD: Berries eaten throughout the range.

MEDICINE: Root bark used by the Meskwaki to expel worms and by the Ojibwa and Missouri River Indians for urinary troubles.
REFERENCES: Densmore 1928:291; Gilmore 1919:32; H. Smith 1928:246, 1932:410; Yanovsky 1936:29.

Ribes cynosbati L.
Prickly gooseberry
Native shrub; occasional in moist and rocky woods in the northern one-half and the southern one-quarter of Illinois.

SYNONYMS: *Grossularia cynosbati* (L.)Mill.
FOOD: Berries eaten fresh, cooked, or preserved by Indians in Minnesota and Wisconsin.
MEDICINE: Root used by the Meskwaki, Potawatomi, and Ojibwa for uterine troubles.
REFERENCES: Densmore 1928:291; H. Smith 1923:71, 1928:246,264, 1932:410, 1933:82,109; Yanovsky 1936:28.

Ribes hirtellum Michx.
Northern gooseberry
Native shrub; occurs in swamps and bogs in northeastern Illinois; rare.

SYNONYMS: *Grossularia hirtella* Michx.
FOOD: Berries eaten by Indians, fresh or dried for storage throughout range.
REFERENCES: H. Smith 1932:410; Yanovsky 1936:28.

Ribes missouriense Nutt.
Missouri gooseberry
Native shrub; common in woods and on riverbanks in the northern two-thirds of Illinois; occasional elsewhere.

SYNONYMS: *Grossularia missouriensis* (Nutt.)Cov. & Britt.
FOOD: Berries eaten by the Missouri River Indians and probably by others as well.
REFERENCES: Gilmore 1919:84; Yanovsky 1936:29.

Saxifraga pensylvanica L.
Swamp saxifrage
Native herb; occasional in spring bogs and moist meadows in the northern one-half of Illinois.

MEDICINE: Used by the Menominee for "medicine."
REFERENCES: H. Smith 1923:53.

SCROPHULARIACEAE
Figwort Family

Castilleja coccinea (L.)Spreng.
Indian paintbrush
Native herb; occasional in prairies and sandy woods in the northern three-fourths of Illinois.

MEDICINE: Root used by the Ojibwa to treat rheumatism and for female troubles.
CEREMONY: Used as a love charm by the Menominee.
REFERENCES: Densmore 1928:288,362; H. Smith 1923:81.

Chelone glabra L.
White turtlehead
Native herb; occasional on wet soil in the northern one-half of Illinois.

MEDICINE: Herb used as a laxative and purgative.
REFERENCES: Tehon 1951:37.

Gerardia flava L.
False foxglove
Native herb; occasional in rocky woods in the southern one-half of Illinois.

SYNONYMS: *Aureolaria flava* (L.)Farw.
MEDICINE: Used by the Sioux as a snakebite remedy.
REFERENCES: Vogel 1970:222.

Gerardia tenuifolia Vahl
False foxglove
Native herb; occasional throughout Illinois on moist soil.

MEDICINE: Herb tea used by the Meskwaki to treat diarrhea.
REFERENCES: H. Smith 1928:246.

Linaria vulgaris Hill
Butter-and-eggs
Introduced herb; occasional throughout Illinois in fields and along roadsides.

MEDICINE: Dried plant used by the Ojibwa as an inhalant in the sweat lodge.
DYE: Recent--flowers can be used for yellow.
REFERENCES: H. Smith 1932:389; Tehon 1951:73.

Melampyrum lineare Lam.
Cow-wheat
Native herb; occurs in bogs, marshes, and dunes in Cook County; rare.

MEDICINE: Infusion used by Ojibwa for eye medicine.
REFERENCES: H. Smith 1932:389.

Mimulus geyeri Torr.
Monkey flower
Native herb; occurs on wet ground in the northern one-half of Illinois; rare.

SYNONYMS: *M. glabratus* H.B.K., *M. glabratus* H.B.K. var. *fremontii* (Benth.)Grant
FOOD: Greens eaten by Indians of New Mexico.
MEDICINE: Leaves used by the Potawatomi for "medicine."
REFERENCES: H. Smith 1933:83; Yanovsky 1936:56.

Pedicularis canadensis L.
Lousewort
Native herb; occasional throughout Illinois on sandy soil in woods and prairies.

FOOD: Greens eaten by the Iroquois.

MEDICINE: Used by the Ojibwa as an aphrodisiac, by the Meskwaki to treat swellings, and by the Potawatomi as a physic and for external swellings.
CEREMONY: Root used by the Ojibwa as a love charm.
REFERENCES: H. Smith 1923:81, 1928:247, 1932:390,432, 1933:83; Waugh 1916:118; Yanovsky 1936:57.

Pedicularis lanceolata Michx.
Swamp wood betony
Native herb; occasional in calcareous fens and wet meadows in the northern one-half of Illinois; uncommon elsewhere.

FOOD: Greens eaten by the Iroquois.
REFERENCES: Waugh 1916:118; Yanovsky 1936:57.

Scrophularia marilandica L.
Late figwort
Native herb; occasional throughout Illinois in rich woods.

MEDICINE: Roots used by the Meskwaki for "medicine."
REFERENCES: H. Smith 1928:247; Tehon 1951:104.

Verbascum thapsus L.
Woolly mullein
Introduced herb; common throughout Illinois in fields and along roadsides.

MEDICINE: Root smoked by the Menominee for pulmonary disease; root decoction drunk by Creeks for coughs and used by Potawatomi, Ojibwa, and Menominee to treat asthma. Recent--leaves and flowers used as a demulcent; flowers used an an embrocation and for tuberculosis.
SMOKING: Leaves added to smoking mixture by several Indian groups.
REFERENCES: Epstein 1981:105; H. Smith 1923:53, 1932:390, 1933:83; Tehon 1951:116.

Veronica arvensis L.
Corn speedwell
Introduced herb; common throughout Illinois in fields, woods, and waste ground.

MEDICINE: Recent--herb used as a diaphoretic, diuretic, and expectorant.
REFERENCES: Tehon 1951:117.

Veronica officinalis L.
Common speedwell
Introduced herb; occasional in grassy areas and open woods in Cook, Dupage, and Richland counties; rare.

MEDICINE: Recent--herb used as a diaphoretic, diuretic, and stimulant.
REFERENCES: Tehon 1951:116.

Veronicastrum virginicum (L.)Farw.
Culver's-root
Native herb; occasional throughout Illinois in woods and prairies.

SYNONYMS: *Veronica virginicum* L.
MEDICINE: Root used by the Menominee as a physic and reviver; by the Meskwaki to cure constipation and to dissolve kidney stones; by the Seneca as a purgative, and by the Ojibwa for "medicine." Recent--roots used as a laxative, cholagogue and emetic.
REFERENCES: Densmore 1928:294; Lewis and Elvin-Lewis 1977:284; H. Smith 1923:53, 1928:247; Tehon 1951:118.

SOLANACEAE
Nightshade Family

Datura stramonium L.
Jimsonweed
Native herb; common throughout Illinois in waste areas.

MEDICINE: The leaves, unripe fruit, and especially the seeds are poisonous. Plant used by many tribes as an intoxicant and hypnotic; seeds or leaves were used as snakebite poultice; leaves used for itching feet and smoked for shortness of breath. Recent--plant used as an antispasmodic, anodyne, and narcotic.
SMOKING: Leaves added to tobacco or chewed for hallucinogenic effects.
REFERENCES: Epstein 1981:108; Tehon 1951:46; Vogel 1970:326-28.

Nicotiana rustica L.
Wild tobacco
Introduced herb; adventive in waste ground.

SMOKING: Plants cultivated and leaves smoked by many Indian groups.
REFERENCES: Vogel 1970:380-84.

Nicotiana tabacum L.
Tobacco
Introduced herb; cultivated and occasionally escaped.

MEDICINE: Leaves used by the Meskwaki as part of the treatment for peyote poisoning. The plant is poisonous if eaten and carcinogenic.
SMOKING: Smoked by all tribes when available.
REFERENCES: H. Smith 1928:197; Vogel 1970:300-304.

Physalis heterophylla Nees
Ground cherry
Native herb; occasional throughout Illinois in woods.

FOOD: Fruit eaten by the Meskwaki and Indians of the Missouri River region, either raw or made into sauce.
MEDICINE: Root used by the Meskwaki for "medicine."
REFERENCES: Gilmore 1919:113; H. Smith 1928:247,264; Yanovsky 1936:56.

Physalis pubescens L.
Annual ground cherry
Native herb; occasional throughout Illinois in fields and disturbed ground.

FOOD: Fruit eaten by Indians in the eastern and central states.
REFERENCES: Yanovsky 1936:56.

Physalis virginiana Mill.
Ground cherry
Native herb; occasional throughout Illinois in woods and disturbed areas.

FOOD: Ripe fruit eaten by the Meskwaki.
MEDICINE: Herb tea used by the Meskwaki to treat dizziness.
REFERENCES: H. Smith 1928:247,264; Yanovsky 1936:56.

Solanum americanum Mill.
Black nightshade
Native herb; occasional throughout Illinois on riverbanks and disturbed ground.

SYNONYMS: *S. nigrum* L.
FOOD: Ripe fruit eaten by California Indians; occurs in several archaeological sites in the American Bottoms. The plant and unripe fruit of species in this genus are poisonous although the thoroughly ripe fruits of some species, including *S. americanum*, are edible.
MEDICINE: Leaf tea used by the Rappahannocks for insomnia.
REFERENCES: Lewis and Elvin-Lewis 1977:55,169; Johannessen 1981; Yanovsky 1936:50.

Solanum carolinense L.
Horse-nettle
Native herb; common throughout Illinois in open woods and waste areas.

MEDICINE: Recent--the ripe berries and rootstock have been used as a sedative and antispasmodic.
REFERENCES: Tehon 1951:107; Vogel 1970:342.

Solanum dulcamara L.
Bittersweet nightshade
Introduced herb; common in waste areas, thickets, and on boggy ground in the northern one-half of Illinois; occasional elsewhere.

MEDICINE: Recent--the young branches have been used as a sedative and hypnotic, chiefly in muscular rheumatism and chronic bronchial and pulmonary affection; compounds present in the plant have potential value in cancer chemotherapy. All parts of the plant are poisonous.
REFERENCES: Lewis and Elvin-Lewis 1977:136; Tehon 1951:107.

STAPHYLEACEAE
Bladdernut Family

Staphylea trifolia L.
Bladdernut
Native shrub; common throughout Illinois in moist woods.

FOOD: Seeds eaten by Indians in the eastern states.
TECHNOLOGY: Twigs used by the Meskwaki for pipestems and the seeds used in rattles.
REFERENCES: H. Smith 1928:248; Yanovsky 1936:41.

THYMELACEAE
Mezereum Family

Dirca palustris L.
Leatherwood
Native shrub; uncommon to rare in rich shaded woods throughout Illinois.

MEDICINE: Bark tea used by the Menominee as a diuretic and for kidney trouble, by the Potawatomi and Ojibwa as a diuretic, and by the Ojibwa as a physic and for pulmonary trouble. Recent--the bark is an emetic; the inner bark sometimes masticated to relieve toothache. The bark and fruit are probably both poisonous.
TECHNOLOGY: Fiber used by the Ojibwa, Potawatomi, Menominee, and Winnebago for cordage and the weaving of bags.
REFERENCES: Densmore 1928:289; Gilmore 1933:137; Lewis and Elvin-Lewis 1977:39,253; H. Smith 1923:54,76, 1932:390, 1933:85,114; Whitford 1941:11.

TILIACEAE
Linden Family

Tilia americana L.
Linden, basswood
Native tree; occasional to common throughout Illinois in rich woods.

FOOD: Young buds, twigs, and sap eaten by the Ojibwa.
MEDICINE: Flowers and leaves used as a stimulant and sedative.
TECHNOLOGY: Bark fiber used by many tribes for thread, cordage, and the weaving of bags and baskets, tying wigwam framework, sewing mats, fish nets, snowshoes, and ropes.
REFERENCES: Densmore 1928:293,321; Gilmore 1919:50, 1933:136-37; H. Smith 1923:76, 1932:422, 1933:114; Tehon 1951:113; Yanovsky 1936:43.

ULMACEAE
Elm Family

Celtis occidentalis L.
Hackberry
Native tree; common throughout Illinois in low woods.

FOOD: Fruit pounded fine by many tribes and used to flavor meat or mixed with parched corn and fat.
DYE: Recent--the roots can be used for yellow.
REFERENCES: Gilmore 1919:76; Lust 1974:536; H. Smith 1928:265; Yanovsky 1936:19.

Celtis laevigata Willd.
Sugarberry
Native tree; occasional to common throughout Illinois in low woods.

FOOD: Fruit used for food and seasoning, as with *C. occidentalis*.
REFERENCES: Yanovsky 1936:19.

Ulmus americana L.
White elm
Native tree; common throughout Illinois in rich woods and floodplains.

MEDICINE: Bark tea used by the Potawatomi for cramps and diarrhea.
TECHNOLOGY: Bark used by the Ojibwa for roofing winter houses and wood by the Indians of the Missouri River region for house posts and mortars and pestles.
REFERENCES: Gilmore 1919:23, 1933:129; H. Smith 1933:86; Vogel 1970:302–304.

Ulmus rubra Muhl.
Slippery elm
Native tree; common throughout Illinois in woods and disturbed areas.

SYNONYMS: *U. fulva* Michx.
FOOD: Cambium used for food by the Iroquois.
MEDICINE: Cambium used by the Menominee as a physic, by the Potawatomi and Ojibwa for sore throats and as a poultice for boils, by the Meskwaki for cold sores and to make a tea to ease childbirth, and by the Indians of the Missouri River region as a laxative; elsewhere for black or bruised eyes and for urinary tract infections.
TECHNOLOGY: Bark used by the Potawatomi for boxes and baskets; by the Potawatomi and Meskwaki to cover wigwams; fiber used for rope and cords.
REFERENCES: Densmore 1928:294; Gilmore 1919:24; Lewis and Elvin-Lewis 1977:224,316; H. Smith 1923:56,77, 1928:252,270, 1932:423,392, 1933:86.

Ulmus thomasi Sarg.
Rock elm
Native tree; occurs in rich woods in the northern one-half of Illinois, uncommon.

TECHNOLOGY: Wood used for bows by the Indians of the Missouri River region.
REFERENCES: Gilmore 1919:23.

UMBELLIFERAE
Carrot Family

Angelica atropurpurea L.
Angelica
Native herb; occasional in woodlands, thickets, and marshes in the northern one-half of Illinois.

MEDICINE: Root used by the Menominee to reduce swellings and by the Meskwaki to treat womb injuries. The plant may cause dermatitis in some persons.
REFERENCES: Lewis and Elvin-Lewis 1977:81; H. Smith 1923:55, 1928:224; Tehon 1951:20; Vogel 1970:272–73.

Carum carvi L.
Caraway
Introduced herb; cultivated and occasionally escaped.

MEDICINE: Root used by the Meskwaki for "female" troubles.
REFERENCES: H. Smith 1928:229.

Cicuta maculata L.
Water hemlock
Native herb; occasional to common throughout Illinois in marshes, wet prairies, and moist woods.

MEDICINE: All parts of the plant, especially the root, are extremely poisonous. The plant was used for "medicine" by the Ojibwa. Recent--the roots and seeds were sometimes used to counteract barbital poisoning.
CEREMONY: The root was smoked by the Ojibwa to attract deer.
REFERENCES: Densmore 1928:288; Lewis and Elvin-Lewis 1977:49; H. Smith 1932:390; Tehon 1951:39; Vogel 1970:241,243.

Conium maculatum L.
Poison hemlock
Introduced herb; occasional throughout Illinois in fields and waste ground.

MEDICINE: All parts, especially the leaves, roots, and unripe fruit are poisonous. Fruit and leaves were used in the past as an antispasmodic, sedative, and anodyne.
REFERENCES: Lewis and Elvin-Lewis 1977:49-50; Tehon 1951:42.

Daucus carota L.
Wild carrot
Introduced herb; common throughout Illinois in fields and waste ground.

FOOD: Recent--the seeds have been used as a coffee substitute.
MEDICINE: Recent--the seeds have been used as a stimulant, diuretic, and aromatic. The plant causes photodermatitis in some persons.
REFERENCES: Lewis and Elvin-Lewis 1977:81,218; Tehon 1951:47.

Eryngium yuccifolium Michx.
Rattlesnake master
Native herb; occasional throughout Illinois in prairies and openings in woods.

MEDICINE: Root used by the Meskwaki as rattlesnake medicine; fruit and leaves used for bladder trouble and to counteract other poisons. Recent--the root was used as a diaphoretic, diuretic, expectorant, and emetic.
REFERENCES: H. Smith 1928:248; Tehon 1951:52; Vogel 1970:103,371.

Heracleum maximum Bartr.
Cow parsnip
Native herb; occasional in low or rich woods in the northern two-thirds of Illinois; rare elsewhere.

SYNONYMS: *H. lanatum* Michx.
FOOD: Buds, stems, and roots eaten by the Meskwaki and in the Northwest.
MEDICINE: Roots used by the Ojibwa as a poultice for sores and boils, sore throats, and indigestion, used by the Meskwaki for stomach cramps, and by the Indians of the Missouri River region for fainting, convulsions, and as a poultice.
CEREMONY: Root or seeds burned by the Ojibwa as a hunting or fishing charm or as evil magic.
REFERENCES: Densmore 1928:289; Gilmore 1919:55; H. Smith 1923:55,82, 1928:249,265, 1932:390; Yanovsky 1936:48.

Osmorhiza claytoni (Michx.)Clarke
Sweet cicely
Native herb; occasional throughout Illinois in rich woods.

MEDICINE: Roots eaten by the Menominee to gain weight; used by the Ojibwa for ulcers and sore throats, tea used to ease childbirth.
REFERENCES: Densmore 1928:290; H. Smith 1923:55,72, 1932:391, 1933:137,304.

Osmorhiza longistylis (Torr.)DC.
Anise-root
Native herb; occasional throughout Illinois in rich woods.

SYNONYMS: *Washingtonia longistylis* (Torr.)Britt.
MEDICINE: Root tea used by the Ojibwa to ease childbirth and for sore throats and amenorrhea, by the Potawatomi as an eye lotion and stomachic, and by the Meskwaki for internal trouble; roots used by the Indians of the Missouri River region as a poultice.
SMOKING: Possibly added to a smoking mixture by the Arikara.
REFERENCES: Epstein 1981:126; Gilmore 1919:55, 1933:138; H. Smith 1928:195, 1932:391, 1933:86; Vogel 1970:137,304.

Pastinaca sativa L.
Wild parsnip
Introduced herb; common in waste areas throughout Illinois.

MEDICINE: Root used by the Potawatomi as a poultice and by the Ojibwa for female troubles. The plant causes photodermatitis in some persons.
REFERENCES: Lewis and Elvin-Lewis 1977:81; H. Smith 1932:91, 1933:86.

Polytaenia nuttallii DC.
Prairie parsley
Native herb; occasional throughout Illinois in prairies and rocky woods.

MEDICINE: Used by the Meskwaki for diarrhea.
REFERENCES: H. Smith 1928:249.

Sanicula canadensis L.
Canadian black snakeroot
Native herb; occasional throughout Illinois in woods.

MEDICINE: Root used by the Ojibwa for "female troubles."
REFERENCES: Densmore 1928:292,360; Tehon 1951:102.

Sanicula gregaria Bicknell
Common snakeroot
Native herb; common throughout Illinois in woods.

MEDICINE: Used by the Meskwaki as an astringent and to stop nosebleeds.
CEREMONY: Possibly used by Menominee sorcerers.
REFERENCES: H. Smith 1923:55, 1928:250.

Sanicula marilandica L.
Black snakeroot
Native herb; occasional on wooded slopes in the northern two-thirds of Illinois; rare elsewhere.

MEDICINE: Root used by the Ojibwa as a poultice for snakebites. Recent--the root was used as an astringent, anodyne, and nervine.
CEREMONY: Possibly used by Menominee sorcerers.
REFERENCES: H. Smith 1923:55, 1932:391; Vogel 1970:212.

Sium suave Walt.
Water parsnip
Native herb; occasional throughout Illinois in marshes, ponds, low ground, and swamps.

FOOD: Although the root is supposedly edible, the plant is suspected of causing livestock poisoning.
CEREMONY: Seeds burned by the Ojibwa as a hunting charm to repel evil spirits.
REFERENCES: Lewis and Elvin-Lewis 1977:50; H. Smith 1932:432.

Taenidia integerrima (L.)Drude
Yellow pimpernel
Native herb; occasional throughout Illinois in prairies and dry, open, rocky woods.

MEDICINE: Root used by the Menominee as a seasoning for other medicines, chewed for bronchial troubles, and made into a tea for pulmonary problems.
SMOKING: Seeds smoked by the Ojibwa for good luck when hunting.
REFERENCES: H. Smith 1923:56, 1932:432.

Thaspium barbinode (Michx.)Nutt.
Hairy meadow parsnip
Native herb; occasional in moist woods in the northern one-half of Illinois.

MEDICINE: Root used by the Ojibwa for colic.
REFERENCES: Densmore 1928:293.

Zizia aurea (L.)Koch
Golden alexanders
Native herb; occasional in moist woods and prairies in the northern three-fourths of Illinois; local elsewhere.

MEDICINE: Root used by the Meskwaki for fevers.
REFERENCES: Gilmore 1919:250; Vogel 1970:175,338.

URTICACEAE
Nettle Family

Boehmeria cylindrica L.
False nettle
Native herb; common throughout Illinois in moist woods.

TECHNOLOGY: Fiber used by the Ojibwa for bowstrings.
REFERENCES: Whitford 1941:12.

Laportea canadensis (L.)Wedd.
Wood nettle
Native herb; common throughout Illinois in moist soil.

MEDICINE: Root used by the Ojibwa, Meskwaki, and Menominee as a diuretic. The plant causes dermatitis in some persons.
TECHNOLOGY: Fiber used by the Ojibwa, Meskwaki, and Menominee for sewing, making twine, and weaving bags.
REFERENCES: Densmore 1928:294; H. Smith 1923:56,77, 1928:250,270, 1932:423,391.

Parietaria pennsylvanica Muhl.
Pellitory
Native herb; common throughout Illinois in shady locations.

MEDICINE: Recent--the herb was used as a diuretic, emmenagogue, and deobstruent.
REFERENCES: Tehon 1951:84.

Urtica dioica L.
Stinging nettle
Native herb; occasional to common in rich woods and moist waste ground in the northern one-half of Illinois; rare elsewhere.

SYNONYMS: *U. gracilis* Ait.
FOOD: Greens of a similar species were eaten by the Iroquois.
MEDICINE: Root used by the Ojibwa for urinary and digestive problems; leaf tea used by the Potawatomi for intermittent fevers. Recent--the herb was used as a diuretic and counterirritant; an extract has long been used in Europe as a hair rinse. The plant is covered with stinging hairs which cause contact dermatitis.
TECHNOLOGY: The fiber was widely used by Indians for sewing, making twine, and weaving bags.
DYE: Recent--the herb may be used for a greenish-yellow.
REFERENCES: Densmore 1928:294; Lewis and Elvin-Lewis 1977:78,313,340; H. Smith 1923:77, 1932:423, 1933:115; Tehon 1951:115; Waugh 1916:118; Yanovsky 1936:20; Whitford 1941:13.

VALERIANACEAE
Valerian Family

Valeriana ciliata T&G.
Valerian
Native herb; occasional in wet prairies and fens in the northern one-fifth of Illinois.

SYNONYMS: *V. edulis* Nutt. ssp. *ciliata* (T&G)F.G.Mey.
MEDICINE: Root used by the Menominee as a poultice for wounds and as a tapeworm medicine and by the Meskwaki to stop hemorrhage.
REFERENCES: Gilmore 1933:57; H. Smith 1928:251.

VERBENACEAE
Vervain Family

Verbena hastata L.
Blue vervain
Native herb; common throughout Illinois in wet wood, wet prairies, and waste ground.

FOOD: Leaves used to make a tea by the Omaha; seeds eaten in California.

MEDICINE: A root tea used by the Menominee to clear cloudy urine; flowers used by the Ojibwa for nosebleeds and by the Indians of the Missouri River region for stomachaches. Recent--the herb was used as a tonic, expectorant, and emetic.
REFERENCES: Densmore 1928:294; Gilmore 1919:22,59; H. Smith 1923:58; Tehon 1951:117; Yanovsky 1936:54.

Verbena urticifolia L.
White vervain
Native herb; occurs in fields, thickets, and disturbed woods in northeastern Illinois; rare.

MEDICINE: Herb used by the Meskwaki to stop profuse menstruation and as a reviver.
REFERENCES: H. Smith 1928:251.

VIOLACEAE
Violet Family

Viola conspersa Reichenb.
Dog violet
Native herb; occurs in moist woods in Cook, Lake and Richland counties; rare.

MEDICINE: Herb tea used by the Ojibwa for heart trouble and skin diseases.
REFERENCES: H. Smith 1932:392; Vogel 1970:386-87.

Viola cucullata Ait.
Marsh blue violet
Native herb; occasional in moist soil in the northern one-half of Illinois.

MEDICINE: Herb used by the Ute Indians for medicine.
REFERENCES: Vogel 1970:386-87.

Viola pedata L.
Birdfoot violet
Native herb; occasional throughout Illinois on cherty slopes and in forest borders.

MEDICINE: Recent--mucilaginous plant used as a demulcent and expectorant; leaves and rhizomes used as a cathartic, emetic, and expectorant.
REFERENCES: Tehon 1951:119; Vogel 1970:386-87.

Viola pubescens Ait.
Downy yellow violet
Native herb; occurs in rich woods in the northern one-half of Illinois; rare.

MEDICINE: The root was used by the Potawatomi to treat heart disease.
REFERENCES: H. Smith 1933:87; Vogel 1970:386-87.

VITACEAE
Grape Family

Parthenocissus quinquefolia (L.)Planch.
Virginia creeper
Native vine; common throughout Illinois in woods.

SYNONYMS: *Psedera quinquefolia* (L.)Greene
FOOD: Fruit and peeled stalks were eaten by the Ojibwa; however, the berries have now been suspected in the poisoning of children.
MEDICINE: Root tea used by the Meskwaki for diarrhea.
DYE: Recent--the fruit may be used for pink.
REFERENCES: Densmore 1928:320; Lust 1974:550; H. Smith 1928:252, 1932:411; Yanovsky 1936:42.

Vitis aestivalis Michx.
Summer grape
Native vine; occasional to common throughout Illinois in rocky woods and on blufftops.

FOOD: Many similar species were used for food by Historic Indian groups.
REFERENCES: Yanovsky 1936:42-43.

Vitis cinerea Engelm.
Winter grape
Native vine; occasional in woods and along streambanks in the southern three-fourths of Illinois.

FOOD: The fruit was eaten by the Indians of the Missouri River region, either fresh or dried, and sap used as a beverage.
REFERENCES: Gilmore 1919:102; Yanovsky 1936:43.

Vitis palmata Vahl
Catbird grape
Native vine; occasional in swamps and low woods in the southern one-sixth of Illinois.

FOOD: Many similar species used for food by Historic Indian groups.
REFERENCES: Yanovsky 1936:42-43.

Vitis riparia Michx.
Frost grape
Native vine, common throughout Illinois on alluvial soil.

FOOD: Many similar species used for food by Historic Indian groups.
REFERENCES: Yanovsky 1936:42-43.

Vitis rupestris Scheele
River-bank grape
Native vine; occurs in rocky woods in Union County; rare.

FOOD: Many similar species used for food by Historic Indian groups.
REFERENCES: Yanovsky 1936:42-43.

Vitis spp.
Grape
Native vine; common throughout Illinois in numerous habitats.

MEDICINE: The roots and branches used by the Ojibwa in a remedy for pulmonary troubles; the sap used as a hair tonic.
TECHNOLOGY: The stems were used for cordage by Indians of the southern states.
REFERENCES: Gilmore 1933:136, Swanton 1946:244.

Vitis vulpina L.
Frost grape
Native vine; occasional throughout Illinois in moist woods.

SYNONYMS: *V. cordifolia* Michx.
FOOD: Berries eaten by many tribes, either fresh or dried, and the sap used as a beverage.
MEDICINE: Seeds were used by the Menominee for removing foreign matter from the eye; twig tea used by the Ojibwa to clear afterbirth and by the Meskwaki to cure insanity and for children who had eaten Indian turnip; the sap was used by the Ojibwa for stomach trouble; elsewhere the root was used for a mouthwash to treat thrush.
REFERENCES: Gilmore 1919:102; Lewis and Elvin-Lewis 1977:263; H. Smith 1923:58,72, 1928:252, 265, 1932:392,411.

Appendix

NATURAL COMMUNITIES OF ILLINOIS DESCRIPTIONS AND LISTS OF POTENTIAL FOOD PLANTS

A *community* is defined as an aggregation of plants and animals having mutual relationships among themselves and to their environment (Oosting 1956:17). It is generally named for its most abundant and conspicuous members and serves as the basic vegetational unit of the plant ecologist. White and Madany (1978:321) note that a *natural community* differs from a plant community in that it is based not solely on vegetation but on all natural features, including vegetation. They further note that natural communities are the smallest units that can be mapped using Illinois Natural Areas inventory techniques, including examination of physiognomy, soil structure, substrate, soil reaction, species composition, vegetation structure, and topographic position (White and Madany 1978:316).

White and Madany (1978) separate the environments of Illinois into nine classes, 28 subclasses, and 93 natural communities (Table 9). Many of the communities occur in more than one of the Natural Divisions defined by Schwegman *et al.* (1973). While stands of a certain community from different parts of the state are more similar to one another than to other communities, there are major differences in structure and composition based on local environmental factors.

The delineation of Natural Divisions (Schwegman *et al.* 1973) and natural communities (White and Madany 1978) has great potential value for archaeologists and other researchers attempting to describe the environment or potential resource availability in a particular area. By identifying the communities adjacent to an archaeological site according to a standard, widely used classification system, it is easier to take advantage of the enormous body of published data already collected by ecologists working throughout Illinois. At the same time, the nature of the vegetation in the archaeological study area is more easily understood by others familiar with the classification.

While a few minor communities might be hard to place in White and Madany's (1978) classification, the majority of environments in the state fall into one of the described communities or are intermediate between such communities. For the purposes of understanding prehistoric plant usage, it does not matter whether a stand is precisely located along a moisture gradient. An upland forest may overall be mesic but will probably contain certain areas that might be dry-mesic or wet-mesic. Such a community could be named *Dry-mesic upland forest/mesic upland forest/wet-mesic upland forest.* Variations in species composition between such closely related communities is slight; and for the purposes of reconstructing prehistoric potential economic plant availability, it is probably adequate, and perhaps more realistic, to be more general in delineating communities.

As noted above, the structure and composition of a community is not identical from one region of the state to another; rather differences occur which reflect variations in environmental factors such as soils and climate. Some species occur, for example, in mesic forest in southern Illinois that do not occur in mesic forest in northern Illinois and *vice versa.* The closer the geographical proximity, the more similar stands of a single community type are likely to be. As distance increases, it becomes necessary to verify the presence of specific plants in a given area by consulting general botanical references such as Mohlenbrock and Ladd (1978) or Jones and Fuller (1955).

The following section lists the natural communities of Illinois including a brief description of each, where they occur, and both dominant and characteristic plants. Dominant plants are certain to be the most common in a natural community; characteristic plants may be common or rare, but their distribution is distinctive for that specific community. Plants with potential economic value are indicated by an asterisk (*); further descriptions of these plants can be found in the section on plant usage. Scientific names are included where the common name is not distinctive. Much of the following information, excluding information of potential food plants, has been abbreviated from White and Madany (1978) which not only contains lengthier descriptions of the various communities but references for the flora and fauna of each community as well.

TABLE 9. Natural Community Classification

CLASS: SUBCLASS	COMMUNITY
FOREST:	
Upland forest	Xeric upland forest
	Dry upland forest
	Dry-mesic upland forest
	Mesic upland forest
	Wet-mesic upland forest
Sand forest	Dry sand forest
	Dry-mesic sand forest
	Mesic sand forest
Floodplain forest	Mesic floodplain forest
	Wet-mesic floodplain forest
	Wet floodplain forest
Flatwoods	Northern flatwoods
	Southern flatwoods
	Sand flatwoods
PRAIRIE:	
Prairie	Dry prairie
	Dry-mesic prairie
	Mesic prairie
	Wet-mesic prairie
	Wet prairie
Sand prairie	Dry sand prairie
	Dry-mesic sand prairie
	Mesic sand prairie
	Wet-mesic sand prairie
	Wet sand prairie
Gravel prairie	Dry gravel prairie
	Dry-mesic gravel prairie
	Mesic gravel prairie
Dolomite prairie	Dry dolomite prairie
	Dry-mesic dolomite prairie
	Mesic dolomite prairie
	Wet-mesic dolomite prairie
	Wet dolomite prairie
Hill prairie	Loess hill prairie
	Glacial drift hill prairie
	Gravel hill prairie
	Sand hill prairie
Shrub prairie	Shrub prairie
SAVANNA:	
Savanna	Dry-mesic savanna
	Mesic savanna
Sand savanna	Dry sand savanna
	Dry-mesic sand savanna
Barren	Dry barren
	Dry-mesic barren
	Mesic barren

TABLE 9. (Concluded)

CLASS: SUBCLASS	COMMUNITY
WETLAND:	
Marsh	Marsh
	Brackish marsh
Swamp	Swamp
	Shrub swamp
Bog	Graminoid bog
	Low shrub bog
	Tall shrub bog
	Forested bog
Fen	Calcareous floating mat
	Graminoid fen
	Low shrub fen
	Tall shrub fen
	Forested fen
Sedge meadow	Sedge meadow
Panne	Panne
Seep & spring	Seep
	Acid gravel seep
	Calcareous seep
	Sand seep
	Spring
LAKE & POND:	
Pond	Pond
Lake	Lake
	Great Lake
STREAM:	
Creek	Low-gradient creek
	Medium-gradient creek
	High-gradient creek
River	Low-gradient river
	Medium-gradient river
	Major stream
PRIMARY:	
Glade	Sandstone glade
	Limestone glade
	Shale glade
Cliff	Sandstone cliff
	Limestone cliff
	Dolomite cliff
	Sandstone overhang
	Eroding bluff community
Lake shore	Beach
	Foredune
CULTURAL:	
Disturbance	Disturbance

Source: Based on White and Madany (1978).

Note: Cave and modern cultural communities have been deleted and a general "disturbance" community added.

FOREST

Communities dominated by trees with an average canopy cover of at least 80% (White and Madany 1978:322).

UPLAND FOREST

Forest communities on areas that do not flood.

XERIC UPLAND FOREST

Description: Open scrubby forest on shallow soil overlying gravel or bedrock; intermediate between dry barren and glade.

Distribution: Shawnee Hills and Cretaceous Hills

Dominant species: *blackjack oak, *post oak, farkleberry.

Characteristic species: *low-bush blueberry.

DRY UPLAND FOREST

Description: A "dry" forest comprised of trees with relatively low water requirements occurs throughout Illinois. Such dry forests are found on exposed south- and west-facing slopes, droughty soils on moraines and other rocky or gravelly glacial features, and on thinly soiled limestone and sandstone ridgetops. The tree species are limited to those which tolerate dry conditions, particularly oaks and hickories, and the total number of species is small compared to the forests of more mesic sites. However, trees are not as stunted as those in xeric upland forest. Sites with fine or "heavy" soils are dominated by post oak and blackjack oak; other areas are usually dominated by black and white oaks and shagbark hickory.

Distribution: Steep ridges at crests of bluffs, and edges of escarpments, throughout the state but especially along the Mississippi River and in the Shawnee Hills.

Dominant species: *shingle oak, *bur oak, *blackjack oak, *rock chestnut oak, *post oak, *black oak.

Characteristic species: *pignut hickory, *black hickory.

Potential food plants: (Table 10:1–2); potential food plants are listed for two areas based on Mohlenbrock's (1967) study of Lake Murphysboro State Park, Jackson County, in the Mt. Vernon Hill Country Division (Schwegman *et al.* 1973) and Wunderlin's (1966) study of Mississippi Palisades State Park, Carroll County, in the Driftless Area Division.

DRY-MESIC UPLAND FOREST

Description: Intermediate between dry and mesic forest.

Distribution: The most prevalent forest type in Illinois; occurs on slopes.

Dominant species: *white oak, *red oak, *black oak.

Characteristic species: *shagbark hickory, *mockernut hickory, *flowering dogwood, *hop hornbeam, black haw.

Potential food plants: (Table 10: 3–5); potential food plants are listed for Mississippi Palisades State Park, Carroll County, in the Driftless Area Division (Wunderlin 1966), Lake Murphysboro State Park, Jackson County, in the Mt. Vernon Hill Country Division (Mohlenbrock 1967), and Cedar Lake Reservoir, Jackson County, in the Shawnee Hills Division (Heineke 1978).

MESIC UPLAND FOREST

Description: Mesic forests generally occupy areas with deep soils, good moisture retention, or slopes and ravines which are protected from sun and wind dessication. The species diversity found in mesic forests is generally greater than that of dry upland forests. In northern Illinois, upland mesic forests are dominated by sugar maple, basswood, and red oak while those of southern Illinois also contain beech, tuliptree, Ohio buckeye, and other species.

Prairie groves, a distinctive feature of the early Illinois landscape, were isolated groves of mesophytic trees. The forests are dominated by sugar maple, red oak, hackberry, bur oak, white ash, white oak, American elm, slippery elm, and basswood, the

composition depending on soil and moisture factors. Prairie groves occur in the Grand Prairie Section of the Grand Prairie Division (Schwegman *et al.* 1973).
Distribution: Throughout Illinois on protected slopes and in ravines.
Dominant species: *sugar maple, *beech, *red oak, *basswood
Characteristic species: *pawpaw, *buckeye, *blue beech, *bitternut hickory, *red mulberry, *bladdernut.
Potential food plants: (Table 10:6–11); potential food plants are given for prairie groves (Funk Forest, Trelease Woods and Brownfield Woods)(Calef 1953), Foley's Woods, Edgar County (Hudnut 1952), slope forest in east-central Illinois (Phillippe and Ebinger 1977), mesic ravines at the Cedar Lake Reservoir in Jackson County in the Shawnee Hills Division (Heineke 1978), and southern Wisconsin (Curtis 1959, Tables VI-1, VI-3).

WET-MESIC UPLAND FOREST

Description: Unusual community on level, poorly drained topography.
Distribution: Small stands throughout the undissected upland forests regions of the state.
Characteristic species: *American elm, *slippery elm, *hackberry, *bur oak.

SAND FOREST

Forest on sand deposits protected from burning.

DRY SAND FOREST

Description: Open scrubby forest on tops of dunes.
Distribution: On sand deposits.
Dominant species: *black oak.
Characteristic species: *black hickory, *blackjack oak.
Potential food plants: See sand savanna.

DRY-MESIC SAND FOREST

Description: Forest on areas with higher soil moisture than dry-sand forest.
Distribution: On sand deposits.
Dominant species: *white oak, *black oak.
Potential food plants: See sand savanna.

MESIC SAND FOREST

Description: Forest on north- and east-facing slopes in sand deposits.
Distribution: Rare, primarily on slopes of sandy river terraces.
Dominant species: *red oak, *white oak, *sugar maple.
Potential food plants: See sand savanna.

FLOODPLAIN FOREST

Floodplain forests occur on alluvial deposits adjacent to stream channels which flood periodically. Species composition of floodplain forests is determined by tolerance to flooding with more tolerant species growing in areas inundated for the longest periods. Drier portions of the floodplain often contain mesic species as well and are somewhat transitional between the wetter floodplain forest and the mesic terrace and slope forests.

MESIC FLOODPLAIN FOREST

Description: Floodplain forest on moderately well-drained, coarse-textured or relatively high elevation soils.
Distribution: Throughout Illinois.
Dominant species: *sugar maple, *white oak, *bur oak, *American elm, *slippery elm, *basswood.
Characteristic species: *black walnut, *white ash.

WET-MESIC FLOODPLAIN FOREST

Description: Floodplain forest intermediate between mesic and wet.

Distribution: Most common floodplain forest community in Illinois, along rivers and creeks throughout the state.

Characteristic species: *silver maple, *hackberry, *sweetgum, *cherry-bark oak, *bur oak, *pin oak, *American elm, *spicebush, *shellbark hickory, *red ash.

Potential food plants: (Table 10: 12–13) Possible food plants are listed for upland streamside forest at Allerton Park, on the Sangamon River in Piatt County (Calef 1953), Lake Murphysboro State Park in the Mt. Vernon Hill Country (Mohlenbrock 1967), and the upper portion of the Embarras River in Douglas County (Phillippe and Ebinger 1977).

WET FLOODPLAIN FOREST

Description: Understory and overstory open due to frequent or prolonged flooding.

Distribution: Throughout Illinois.

Characteristic species: *silver maple, *cottonwood, *sycamore, *red maple, *river birch, *black willow, *boxelder.

Potential food plants: (Table 10:14–18) Possible food plants are listed for the upper portion of the Embarras River (Crites and Ebinger 1969), Southern Illinois rivers and streams (Mohlenbrock *et al.* 1961), southern Illinois floodplains (Mohlenbrock 1967), the Sangamon River Basin (A. Jones and Bell 1974:3–4), and southern Wisconsin (Curtis 1959, Tables VII-1 and VII-3).

FLATWOODS

Upland flatwoods are relatively rare in Illinois. They occur on flat-lying areas with poor surface and subsurface drainage and fine-grained soils. Poor soil aeration and the lack of oxygen reaching the tree roots is a critical factor in limiting the species which grow in these situations.

NORTHERN FLATWOODS

Description: Forest on level or nearly level soil with a perched water table; covered by savanna vegetation prior to Euro-American settlement.

Distribution: Poorly drained uplands on the Valparaiso moraine.

Dominant species: *white oak, *swamp white oak, *shingle oak, *American elm.

Characteristic species: winterberry, sedge.

SOUTHERN FLATWOODS

Description: Stunted forest found on level areas with well-developed hardpan, usually on Illinoian-age till.

Distribution: Southern Illinois in the area of Illinoian glaciation.

Dominant species: *post oak, *blackjack oak, *pin oak, *swamp white oak, *white oak, *spanish oak.

Characteristic species: wood reed.

Potential food plants: (Table 10:19); list compiled from Voigt and Mohlenbrock (1964:116–118).

SAND FLATWOODS

Description: Forest developed on acid, peaty sand overlying clay.

Distribution: Sandy plains in northern Illinois.

Dominant species: *pin oak, *white oak, *sour gum, *red maple.

Characteristic species: winterberry, *false lily-of-the-valley, *partridge-berry, *cinnamon fern, *low-bush blueberry.

TABLE 10. Potential Food Plants of Forest Communities

SCIENTIFIC NAME	COMMON NAME	USE*	1** Dry Southern Ill. Ridgetops	2 Dry Driftless Area Uplands	3 Dry-Mesic Driftless Area	4 Dry-Mesic Southern Ill. Uplands	5 Dry-Mesic Shawnee Hills Uplands	6 Mesic Prairie Groves	7 Mesic Eastern Ill. Uplands	8 Mesic Eastern Ill. Uplands	9 Mesic Shawnee Hills Ravines	10 Mesic Southern Wisc. Uplands	11 Mesic East-central Ill. Uplands	12 Central Ill. Streamside	13 Southern Ill. Streamside	14 East-Central Ill. Floodplain	15 Southwest Ill. Floodplain	16 Southern Ill. Floodplain	17 Central Ill. Floodplain	18 Southern Wisc. Floodplain	19 Southern Ill. Flatwoods	TOTAL
ACER NEGUNDO	Boxelder	C2/B						*						*	*	*	*	*	*	*		8
A. RUBRUM	Red Maple	C2/B				*	*				*	*			*		*					6
A. SACCHARUM	Sugar Maple	C2/B	*		*			*	*	*	*	*	*	*			*		*		*	12
A. SACCHARINUM	Silver Maple	C2/B						*	*					*	*	*	*	*	*	*	*	10
ALLIUM CANADENSE	Wild Garlic	T28						*						*			*		*			4
A. TRIOCCUM	Wild Leek	T28										*	*	*								3
AMELANCHIER ARBOREA	Shadbush	F4	*			*	*															3
AMPHICARPA BRACTEATA	Hog Peanut	F28			*	*	*	*				*				*			*	*		8
APIOS AMERICANA	Groundnut	T9													*				*	*		3
ARALIA NUDICAULIS	Wild Sarsaparilla	TE			*																	1
A. RACEMOSA	American Spikenard	T28/G3			*			*			*											3
ARISAEMA TRIPHYLLUM	Jack-in-the-Pulpit	T28			*			*	*	*	*	*	*	*			*		*		*	11
ARUNDINARIA GIGANTEA	Giant Cane	S4									*						*	*				3
ASCLEPIAS INCARNATA	Swamp Milkweed	G345																	*			1
A. SYRIACA	Common Milkweed	G345						*						*								2
ASIMINA TRILOBA	Pawpaw	F7				*		*	*		*			*			*					6
CARYA CORDIFORMIS	Bitternut Hickory	N7	*	*		*	*	*		*		*		*			*		*	*	*	12
C. GLABRA	Pignut Hickory	N7	*			*			*	*	*		*				*				*	8
C. ILLINOENSIS	Pecan	N7									*						*		*		*	4
C. LACINIOSA	Shellbark Hickory	N7							*		*								*	*	*	5
C. OVALIS	Sweet Pignut Hickory	N7				*	*	*			*											4
C. OVATA	Shagbark Hickory	N7		*				*	*	*	*	*	*						*	*	*	10
C. TEXANA	Black Hickory	N7					*															1
C. TOMENTOSA	Mockernut Hickory	N7				*				*	*		*	*							*	6
CELTIS OCCIDENTALIS	Hackberry	F67				*		*	*		*	*		*		*	*			*		9
CLAYTONIA VIRGINICA	Spring Beauty	T3	*			*	*		*	*	*	*	*				*				*	10
CORYLUS AMERICANA	Hazelnut	N6		*		*	*			*			*								*	6
CRATAEGUS MOLLIS	Hawthorn	F67				*		*	*					*								4
CYPERUS ARISTATUS	Plantain-leaved Sedge	T28																*				1
C. ESCULENTUS	Nut-grass	T28																*				1

TABLE 10. (Continued).

SCIENTIFIC NAME	COMMON NAME	USE*	Dry Southern Ill. Ridgetops 1**	Dry Driftless Area Uplands 2	Dry-Mesic Driftless Area 3	Dry-Mesic Southern Ill. Uplands 4	Dry-Mesic Shawnee Hills Uplands 5	Mesic Prairie Groves 6	Mesic Eastern Ill. Uplands 7	Mesic Eastern Ill. Uplands 8	Mesic Shawnee Hills Ravines 9	Mesic Southern Wisc. Uplands 10	Mesic East-central Ill. Uplands 11	Central Ill. Streamside 12	Southern Ill. Streamside 13	East-central Ill. Floodplain 14	Southwest Ill. Floodplain 15	Southern Ill. Floodplain 16	Central Ill. Floodplain 17	Southern Wisc. Floodplain 18	Southern Ill. Flatwoods 19	TOTAL
DENTARIA LACINIATA	Toothwort	T23			*			*	*	*	*	*	*				*				*	9
DICENTRA CANADENSIS	Toothwort	T3						*			*			*								3
DIOSPYROS VIRGINIANA	Persimmon	F7	*			*	*										*				*	5
ELYMUS CANADENSIS	Nodding Wild Rye	S7	*	*		*																3
ERYTHRONIUM ALBIDUM	Dog-tooth Violet	T28	*					*	*	*	*	*		*							*	8
FAGUS GRANDIFOLIA	Beech	N7				*					*	*					*					4
FRAXINUS PENNSYLVANICA	Green Ash	CE					*				*	*			*	*				*		6
GLYCERIA STRIATA	Fowl Manna Grass	S56						*			*						*	*				4
GYMNOCLADUS DIOICA	Kentucky Coffee Tree	S7							*	*		*				*					*	5
HELIANTHUS TUBEROSUS	Jerusalem Artichoke	T278		*																		1
HERACLEUM MAXIMUM	Cow Parsnip	T4/G4												*								1
HYDROPHYLLUM APPENDICULATUM	Great Waterleaf	G3						*				*	*	*								4
H. CANADENSIS	Broad-leaved Waterleaf	TE						*						*								2
H. VIRGINIANUM	Virginia Waterleaf	G3	*		*			*	*		*	*	*	*					*			9
JUGLANS CINEREA	Butternut	N7			*			*			*	*							*			5
J. NIGRA	Black Walnut	N7	*	*	*	*		*	*	*	*	*	*	*		*	*				*	14
LYCOPUS UNIFLORUS	Northern Bugle Weed	T28																		*		1
MALUS CORONARIA	Wild Crabapple	F6				*	*															2
M. IOENSIS	Iowa Crabapple	F6				*																1
MORUS RUBRA	Red Mulberry	F5				*		*	*		*					*			*			6
ONOCLEA SENSIBILIS	Sensitive Fern	T8		*					*		*		*	*					*			6
OXALIS STRICTA	Yellow Wood Sorrel	G4-7	*			*																2
O. VIOLACEA	Violet Wood Sorrel	G4-7/T	*										*				*					3
PARTHENOCISSUS QUINQUEFOLIA	Virginia Creeper	F7	*			*	*		*		*	*		*			*		*	*		10
PHYTOLACCA AMERICANA	Pokeberry	G3	*			*		*	*					*			*					6
PODOPHYLLUM PELTATUM	Mayapple	F6	*	*		*			*	*	*	*	*				*				*	10
POLYGONATUM COMMUTATUM	Solomon's Seal	T28				*				*	*		*	*			*				*	7
POPULUS DELTOIDES	Cottonwood	CE														*	*	*	*	*		5
P. TREMULOIDES	Trembling Aspen	CE																		*		1
PRUNUS AMERICANA	American Plum	F6				*	*															2

TABLE 10. (Continued).

SCIENTIFIC NAME	COMMON NAME	USE	Dry Southern Ill. Ridgetops 1	Dry Driftless Area Uplands 2	Dry-Mesic Driftless Area 3	Dry-Mesic Southern Ill. Uplands 4	Dry-Mesic Shawnee Hills Uplands 5	Mesic Prairie Groves 6	Mesic Eastern Ill. Uplands 7	Mesic Eastern Ill. Uplands 8	Mesic Shawnee Hills Ravines 9	Mesic Southern Wisc. Uplands 10	Mesic East-central Ill. Uplands 11	Central Ill. Streamside 12	Southern Ill. Streamside 13	East-central Ill. Floodplain 14	Southwest Ill. Floodplain 15	Southern Ill. Floodplain 16	Central Ill. Floodplain 17	Southern Wisc. Floodplain 18	Southern Ill. Flatwoods 19	TOTAL
PRUNUS HORTULANA	Wild Goose Plum	F67				*	*															2
PRUNUS SEROTINA	Black Cherry	F56	*			*	*	*			*	*	*	*			*			*		10
QUERCUS ALBA	White Oak	N7	*	*	*	*	*		*	*	*	*	*	*	*		*			*	*	15
Q. BICOLOR	Swamp White Oak	N7										*							*	*	*	4
Q. BOREALIS	Red Oak	N7										*										1
Q. COCCINEA	Scarlet Oak	N7					*															1
Q. ELLIPSOIDALIS	Hill's Oak	N7			*																	1
Q. FALCATA	Spanish Oak	N7				*	*															2
Q. IMBRICARIA	Shingle Oak	N7					*		*								*					3
Q. LYRATA	Overcup Oak	N7																*			*	2
Q. MACROCARPA	Bur Oak	N7		*				*	*	*		*		*		*	*		*	*	*	11
Q. MARILANDICA	Blackjack OaK	N7	*			*																2
Q. MICHAUXII	Basket Oak	N7																*				1
Q. MUHLENBERGII	Yellow Oak	N7	*			*	*	*	*	*							*					7
Q. PALUSTRIS	Pin Oak	N7													*		*	*			*	4
Q. RUBRA	Red Oak	N7	*	*	*	*		*	*	*	*		*	*			*				*	12
Q. SHUMARDII	Shumard's Oak	N7									*											1
Q. STELLATA	Post Oak	N7	*			*																2
Q. VELUTINA	Black Oak	N7	*	*	*	*	*	*	*	*		*	*	*			*			*	*	14
RHUS COPALLINA	Dwarf Sumac	F67	*			*																2
R. GLABRA	Smooth Sumac	F67	*			*																2
RIBES CYNOSBATI	Prickly Gooseberry	F5										*										1
R. MISSOURIENSE	Missouri Gooseberry	F5		*				*	*				*	*								5
RORIPPA ISLANDICA	Marsh Yellow Cress	G3																	*			1
RUBUS ALLEGHENIENSIS	Common Blackberry	F6												*								1
R. ARGUTUS	High-bush Blackberry	F45	*			*																2
R. ENSLENII	Arching Dewberry	F45	*																			1
R. FLAGELLARIS	Dewberry	F56				*																1
R. OCCIDENTALIS	Black Raspberry	F56				*	*	*						*								4
R. PENNSYLVANICUS	Blackberry	F6				*					*											2

TABLE 10. (Concluded)

SCIENTIFIC NAME	COMMON NAME	USE*	1 Dry Southern Ill. Ridgetops	2 Dry Driftless Area Uplands	3 Dry-Mesic Driftless Area	4 Dry-Mesic Southern Ill. Uplands	5 Dry-Mesic Shawnee Hills Uplands	6 Mesic Prairie Groves	7 Mesic Eastern Ill. Uplands	8 Mesic Eastern Ill. Uplands	9 Mesic Shawnee Hills Ravines	10 Mesic Southern Wisc. Uplands	11 Mesic East-central Ill. Uplands	12 Central Ill. Streamside	13 Southern Ill. Streamside	14 East-central Ill. Floodplain	15 Southwest Ill. Floodplain	16 Southern Ill. Floodplain	17 Central Ill. Floodplain	18 Southern Wisc. Floodplain	19 Southern Ill. Flatwoods	TOTAL
RUDBECKIA HIRTA	Black-eyed Susan	G3						*								*			*	*		4
SAGITTARIA LATIFOLIA	Common Arrowhead	T6-8																*				1
SAMBUCUS CANADENSIS	Elderberry	F67	*			*		*	*					*					*	*		7
SMILACINA RACEMOSA	False Solomon's Seal	F5	*			*		*	*		*	*	*	*			*				*	10
SMILACINA STELLATA	False Solomon's Seal	F5						*		*				*								3
SMILAX BONA-NOX	Catbrier	T28					*															1
S. GLAUCA	Catbrier	T28				*	*										*					3
S. ROTUNDIFOLIA	Catbrier	T28	*				*															2
STAPHYLEA TRIFOLIA	Bladdernut	SG7				*		*	*	*	*	*	*	*			*		*		*	11
STROPHOSTYLES HELVOLA	Wild bean	T28				*	*															2
SYMPHORICARPOS OCCIDENTALIS	Wolfberry	F67		*																		
TILIA AMERICANA	Basswood	B2						*	*	*	*	*		*			*		*	*	*	10
ULMUS RUBRA	Slippery Elm	CE	*			*		*	*	*		*	*	*		*	*			*	*	12
URTICA DIOICA	Stinging Nettle	G345						*								*				*		3
UVULARIA SESSIFOLIA	Bellwort	G3															*					1
VACCINIUM VACILLANS	Low-bush Blueberry	F6					*															1
VITIS AESTIVALIS	Summer Grape	F7				*	*	*	*		*							*				6
V. CINEREA	Winter Grape	F7	*			*	*				*							*				5
V. PALMATA	Catbird Grape	F7					*				*											2
V. RIPARIA	Frost Grape	F6									*			*					*	*		4
V. RUPESTRIS	Riverbank Grape	F6									*											1
V. VULPINA	Frost Grape	F7				*					*											2
XANTHIUM COMMUNE	Cocklebur	S7														*						1
TOTAL			31	14	13	49	30	41	33	23	42	31	24	38	7	14	38	13	26	24	29	520

* C=Bark or Cambium, F=Fruit, G=Greens, S=Seeds, N=Nuts, T=Roots or tubers, P=Pollen, 1=Winter, 2=Early Spring, 3=Spring, 4=Early summer, 5=Late Summer, 6=Early Fall, 7=Fall, 8=Late Fall, 9=All Year, E=Emergency

** See Table 3 For References.

PRAIRIE

Because of a unique combination of climate and soils, prairie was, for thousands of years prior to Euro-American settlement, the dominant type of vegetation over much of Illinois. While prairie communities are dominated by grasses, they include a rich diversity of herbaceous plants as well. Although the biomass produced by the prairie each year is tremendous, the majority of plants do not yield potential human food resources.

Prairie communities occur along a moisture gradient stretching from excessively well-drained sand prairies and loess and gravel hill prairies to excessively wet and poorly drained sites which ultimately grade into marshes. Based on soils, topography and moisture, a series of prairie communities have been described for Illinois (White and Madany 1978,1981).

PRAIRIE

DRY PRAIRIE

Description: Prairie less than 1 meter in height on steep, exposed and excessively drained slopes.
Distribution: Rare because such slopes are usually forested.
Dominant species: *little bluestem, sideoats grama, *porcupine grass.
Potential food plants: (Table 11: 20); based on Sampson (1921).

DRY-MESIC PRAIRIE

Description: Prairie attaining greater height and species diversity than the dry prairie.
Distribution: Throughout the prairie region.
Dominant species: *little bluestem, Indian grass, *porcupine grass.
Characteristic species: *leadplant, *pale coneflower, rough blazing-star, *prairie cinquefoil.

MESIC PRAIRIE

Description: Prairie with maximum species diversity and maximum grass and forb height.
Distribution: Originally one of the most widespread and characteristic in Illinois.
Dominant species: *big bluestem, Indian grass, prairie dropseed.
Characteristic species: *cream wild indigo, shooting-star, *rattlesnake master, blazing-star, hoary puccoon, *white prairie clover, *downy phlox, *compass-plant, prairie-dock.
Potential food plants: (Table 11: 21); based on studies by Sampson (1921), Gleason (1910), and A. Jones and Bell (1974).

WET-MESIC PRAIRIE

Description: Prairie characterized by standing water after rains because of high water table.
Distribution: Level areas between wet and mesic prairie.
Dominant species: *big bluestem, bluejoint grass, switch grass, indian grass, *cord grass.
Characteristic species: loosestrife, prairie sundrops, smooth phlox, ragwort, *culver's root, golden alexanders.

WET PRAIRIE

Description: Prairie found on almost constantly saturated soil.
Distribution: Throughout the prairie regions of Illinois.
Dominant species: bluejoint grass, sedge, *cord grass.
Characteristic species: prairie Indian-plantain, *common boneset, *wild blue iris, winged loosestrife, water parsnip.
Potential food plants: (Table 11:22-24); based on Sampson (1921) and Turner (1936).

TABLE 11. Potential Food Plants of Prairie Communities.

SCIENTIFIC NAME	COMMON NAME	USE*	Dry Prairie 20**	Mesic Prairie 21	Wet Prairie 22	Miss. River Floodplain Prairie 23	Ill. River Floodplain Prairie 24	Dry Sand Prairie 25	Wet Sand Prairie 26	Wet Prairie, Lake Michigan 27	Loess hill Prairie 28	Glacial Drift Hill Prairie 29	Gravel Hill Prairie 30	Savanna 31	Dry Sand Savanna 32	Black Oak Savanna 33	Sand Thicket 34	Bur Oak Savanna 35	Mixed Oak Savanna 36	Barrens 37	TOTAL
ACER NEGUNDO	Boxelder	C2/B								*											1
A. SACCHARINUM	Silver Maple	C2/B																	*		1
ALLIUM CANADENSE	Wild Garlic	T28		*																	1
A. STELLATUM	Cliff Onion	T28									*										1
AMARANTHUS HYBRIDUS	Green Amaranth	G3/S				*	*														2
AMPHICARPA BRACTEATA	Hog Peanut	F28										*				*		*			3
APIOS AMERICANA	Groundnut	T9		*	*	*	*									*					5
ARALIA NUDICAULIS	Wild Sarsaparilla	TE														*					1
ARCTOSTAPHYLOS UVA-URSI	Bearberry	F67								*						*					2
ASCLEPIAS INCARNATA	Swamp Milkweed	G345			*	*										*					3
A. SYRIACA	Common Milkweed	G345		*	*				*				*								4
A. TUBEROSA	Butterfly-weed	G345		*							*	*	*	*	*	*			*		8
ASTER MACROPHYLLUS	Big-leaved Aster	G3														*					1
ASTRAGALUS CANADENSIS	Canadian Milk Vetch	T37									*										1
CARYA CORDIFORMIS	Bitternut Hickory	N7														*			*		2
C. OVATA	Shagbark Hickory	N7												*				*		*	3
C. TOMENTOSA	Mockernut Hickory	N7																		*	1
CELASTRUS SCANDENS	Bittersweet	CE										*									1
CELTIS OCCIDENTALIS	Hackberry	F67																	*		1
CORYLUS AMERICANA	Hazelnut	N6										*			*				*		3
CORNUS STOLONIFERA	Red Osier	F6															*				1
CRATAEGUS MOLLIS	Hawthorn	F67											*								1
CYPERUS ARISTATUS	Plantain-leaved Sedge	T28			*																1
ELYMUS CANADENSIS	Nodding Wild Rye	S7		*				*			*		*	*						*	6
EQUISETUM LAEVIGATUM	Smooth Scouring Rush	G5						*	*		*										3
FRAGARIA AMERICANA	Hillside Strawberry	F4																*			1
F. VIRGINIANA	Wild Strawberry	F4		*				*			*	*	*	*		*		*	*		9
FRAXINUS PENNSYLVANICUS	Green Ash	CE														*			*		2
GAYLUSSACIA BACCATA	Black Huckleberry	F67													*						1
GLYCERIA STRIATA	Fowl Manna Grass	S56		*																	1

TABLE 11. (Continued).

SCIENTIFIC NAME	COMMON NAME	USE	Dry Prairie 20	Mesic Prairie 21	Wet Prairie 22	Miss. River Floodplain Prairie 23	Ill. River Floodplain Prairie 24	Dry Sand Prairie 25	Wet Sand Prairie 26	Wet Prairie, Lake Michigan 27	Loess hill Prairie 28	Glacial Drift Hill Prairie 29	Gravel Hill Prairie 30	Savanna 31	Dry Sand Savanna 32	Black Oak Savanna 33	Sand Thicket 34	Bur Oak Savanna 35	Mixed Oak Savanna 36	Barrens 37	TOTAL
GYMNOCLADUS DIOICA	Kentucky Coffee Tree	S7																	*		1
HORDEUM JUBATUM	Squirrel-tail Grass	S56			*								*								2
IVA ANNUA	Sumpweed	S56				*	*														2
JUGLANS NIGRA	Black Walnut	N7														*	*				2
KOELERIA CRISTATA	June Grass	S5	*					*		*	*		*		*	*	*				8
LATHYRUS MARITIMUS	Beach Pea	G3								*											1
L. PALUSTRIS	Marsh Vetchling	S6			*					*							*				3
MAIANTHEMUM CANADENSE	Wild Lily-of-the-Valley	F678														*					1
MALUS IOENSIS	Iowa Crabapple	F6										*	*				*				3
MORUS RUBRA	Red Mulberry	F5												*					*		2
ONOCLEA SENSIBILIS	Sensitive Fern	T8			*																1
OROBANCHE FASCICULATA	Clustered Broomrape	G3						*													1
OXALIS STRICTA	Yellow Wood Sorrel	G4-7								*	*										2
O. VIOLACEA	Violet Wood Sorrel	G4-7		*							*		*								3
PANICUM CAPILLARE	Witch Grass	S67		*		*	*				*		*								5
PARTHENOCISSUS QUINQUEFOLIA	Virginia Creeper	F7														*	*	*	*		4
PEDICULARIS CANADENSIS	Lousewort	G3		*								*				*		*			4
P. LANCEOLATA	Swamp Wood Betony	G3			*												*				2
PETALOSTEMUM CANDIDUM	White Prairie Clover	T?		*				*			*	*	*			*	*			*	8
P. PURPUREUM	Purple Prairie Clover	T?		*				*		*	*	*	*	*		*					8
PHYSALIS HETEROPHYLLA	Ground Cherry	F67	*	*				*			*	*	*		*	*					8
P. PUBESCENS	Annual Ground Cherry	F67									*										1
P. VIRGINIANA	Ground Cherry	F67	*	*				*			*		*			*					6
POLYGONATUM COMMUTATUM	Solomon's Seal	T28											*			*		*			3
POLYGONUM COCCINEUM	Water Smartweed	S56			*																1
POPULUS DELTOIDES	Cottonwood	CE							*								*				2
P. GRANDIDENTATA	Large-toothed Aspen	CE														*		*			2
P. TREMULOIDES	Trembling Aspen	CE							*				*			*		*			4
PRUNUS AMERICANA	American Plum	F6											*								1
P. SEROTINA	Black Cherry	F56												*	*	*	*	*	*		6

TABLE 11. (Concluded).

SCIENTIFIC NAME	COMMON NAME	USE	Dry Prairie 20	Mesic Prairie 21	Wet Prairie 22	Miss. River Floodplain Prairie 23	Ill. River Floodplain Prairie 24	Dry Sand Prairie 25	Wet Sand Prairie 26	Wet Prairie, Lake Michigan 27	Loess hill Prairie 28	Glacial Drift Hill Prairie 29	Gravel Hill Prairie 30	Savanna 31	Dry Sand Savanna 32	Black Oak Savanna 33	Sand Thicket 34	Bur Oak Savanna 35	Mixed Oak Savanna 36	Barrens 37	TOTAL
PRUNUS SUSQUEHANAE	Sand Cherry	F56								*							*				2
P. VIRGINIANA	Common Chokecherry	F67											*			*	*	*	*		5
PTERIDIUM AQUILINUM	Bracken Fern	G3													*			*			2
QUERCUS ALBA	White Oak	N7													*	*		*		*	4
Q. ELLIPSOIDALIS	Hill's Oak	N7														*	*				2
Q. IMBRICARIA	Shingle Oak	N7																		*	1
Q. MACROCARPA	Bur Oak	N7												*		*		*			3
Q. MARILANDICA	Black Oak	N7												*		*			*	*	4
Q. RUBRA	Red Oak	N7														*			*		2
Q. STELLATA	Post Oak	N7																		*	1
Q. VELUTINA	Black Oak	N7								*				*	*	*		*	*		6
RHUS COPALLINA	Dwarf Sumac	F67													*						1
R. GLABRA	Smooth Sumac	F67									*		*	*	*			*	*		6
RUBUS ALLEGHENIENSIS	Common Blackberry	F6											*	*							2
R. FLAGELLARIS	Dewberry	F56											*								1
R. OCCIDENTALIS	Black Raspberry	F56											*			*	*	*	*		5
R. STRIGOSUS	Red Raspberry	F56																*	*		2
SAGITTARIA LATIFOLIA	Common Arrowhead	T6-8			*																1
SAMBUCUS CANADENSIS	Elderberry	F67														*					1
SCIRPUS VALIDUS	Soft-stem Bulrush	T28/P5/S4-7							*												1
SMILACINA RACEMOSA	False Solomon's Seal	F5													*	*	*		*		4
S. STELLATA	False Solomon's Seal	F5													*	*					2
SOLANUM AMERICANUM	Black Nightshade	F67														*			*		2
SPOROBOLUS CRYPTANDRUS	Sand Dropseed	S67						*			*		*								3
STROPHOSTYLES HELVOLA	Wild Bean	T28				*	*	*								*			*		5
SYMPHORICARPOS OCCIDENTALIS	Wolfberry	F67											*								1
ULMUS RUBRA	Slippery Elm	CE												*							1
VACCINIUM ANGUSTIFOLIA	Low-Bush Blueberry	F6													*						1
VERBENA HASTATA	Blue Vervain	S6		*			*			*											3
VIBURNUM LENTAGO	Nannyberry	F67														*					1
VITIS RIPARIA	Frost Grape	F6												*							1
V. VULPINA	Frost Grape	F7														*	*	*			3
VULPIA OCTOFLORA	Six-weeks Fescue	S5						*			*		*								3
TOTAL			3	15	10	6	6	12	5	10	18	10	25	14	14	38	16	19	21	8	250

* C=Bark or Cambium, F=Fruit, G=Greens, S=Seeds, N=Nuts, T=Roots or tubers, P=Pollen, 1=Winter, 2=Early Spring, 3=Spring, 4=Early summer, 5=Late Summer, 6=Early Fall, 7=Fall, 8=Late Fall, 9=All Year, E=Emergency

** See Table 3 For References.

SAND PRAIRIE

Sand prairie originally occupied large areas of glacial outwash sand. The flora is similar to that of dry prairie with the addition of several species more common on the Great Plains such as western ragweed, prickly-pear, poppy mallow, hairy grama, western sunflower, silky aster and flax-leaved aster. The list of potential food plants is based on the study by Gleason (1910). Separate listings are given for wet and dry sand prairie.

DRY SAND PRAIRIE

Description: Prairie characterized by grasses less than 1 meter in height, no soil A horizon.
Distribution: Primarily found on the crests of sand dunes; rare because such locations are generally savanna instead.
Dominant species: *little bluestem, sand reed, *june grass, *cord grass.
Characteristic species: stiff sandwort, beach wormwood, poppy mallow, *horsemint, *prickly-pear.
Potential food plants: (Table 11:25); based on Gleason (1910).

DRY-MESIC SAND PRAIRIE

Description: Prairie characterized by taller and more diverse plants and by a dark A horizon.
Distribution: In association with any other sand prairie.
Dominant species: *little bluestem, Indian grass, cord grass.
Characteristic species: flax-leaved aster, rough blazing-star, *showy goldenrod, *birdfoot violet.

MESIC SAND PRAIRIE

Description: Prairie composed of grasses and forbs as well as mosses and low shrubs; soil has a dark, deep A horizon.
Distribution: Sand areas, most common in northeastern Illinois.
Dominant species: *big bluestem, *little bluestem, Indian grass.
Characteristic species: *colic root, *black chokeberry, purple chokeberry, flat-topped aster, grass pink orchid, downy sunflower, *American feverfew, *swamp dewberry, nut rush, *low-bush blueberry.

WET-MESIC SAND PRAIRIE

Description: Prairie characterized by a deep, dark, acid A horizon and the presence of standing water on the surface for short periods after heavy precipitation.
Distribution: Small areas usually associated with mesic sand prairie.
Dominant species: *big bluestem, bluejoint grass, sedge, Indian grass, *cord grass.
Characteristic species: *royal fern, *mountain-mint, meadow beauty, lance-leaved violet, yellow-eyed grass.

WET SAND PRAIRIE

Description: Very similar to wet prairie; standing water for up to four months per year.
Distribution: Sand areas.
Dominant species: bluejoint grass, sedge, *cord grass, marsh fern.
Potential food plants: (Table 11:26-27); based on Gleason (1910) and Gates (1912).

GRAVEL PRAIRIE

Prairie developed on gravel or very gravelly soils, usually calcareous and well drained.

DRY GRAVEL PRAIRIE

Description: Prairie found on steep gravel slopes; plants usually less than 1 meter in height..
Distribution: Kames and eskers in the Northeastern Morainal Division and on slopes of gravel terraces along rivers.
Dominant species: *little bluestem, sideoats grama.
Characteristic species: *pasque-flower, stiff sandwort, woolly milkweed, wild flax, *yellow puccoon, prairie buttercup, kitten tails.

DRY-MESIC GRAVEL PRAIRIE

Description: Intermediate between dry and mesic gravel prairie.

Distribution: Found in association with dry gravel prairie on relatively moist lower slopes.

Dominant species: *little bluestem, Indian grass, prairie dropseed, *cord grass.

Characteristic species: *Aster ptarmicoides* (aster), *scurf-pea, *small scullcap.

MESIC GRAVEL PRAIRIE

Description: Similar to mesic prairie because of high soil moisture.

Distribution: Found on the lower slopes of valley train deposits in the northeastern morainal division.

Dominant species: *little bluestem, Indian grass, prairie dropseed.

Characteristic species: low calamint, *valerian.

DOLOMITE PRAIRIE

Occurs where dolomite is less than 1.5m below the suface. The soils are shallow and have a high pH.

DRY DOLOMITE PRAIRIE

Description: Short prairie found where soil overlying dolomitic bedrock is shallow or absent.

Distribution: Rock River Hill Country and along the lower Des Plaines and Kankakee rivers.

Dominant species: *little bluestem, sideoats grama.

Characteristic species: pagoda plant, false boneset, *Muhlenbergia cuspidata* (muhly), hairy beard-tongue.

DRY-MESIC DOLOMITE PRAIRIE

Description: Taller and more diverse than the dry dolomite prairie with which it is generally associated.

Distribution: Found on slightly deeper soils and lower slopes.

Dominant species: *little bluestem, Indian grass, *cord grass.

MESIC DOLOMITE PRAIRIE

Description: Dolomite prairie found where soil depths equal 15 cm or greater; similar to mesic prairie but with a slightly different species composition.

Distribution: Found along the lower Des Plaines and Kankakee river valleys, perhaps originally occurring elsewhere in northern Illinois.

Dominant species: *big bluestem, Indian grass, prairie dropseed.

Characteristic species: northern bedstraw, leafy prairie clover.

WET-MESIC DOLOMITE PRAIRIE

Description: Prairie found where soil overlying dolomite averages 30 cm and there is standing water for short periods after heavy precipitation; very similar to graminoid fen.

Distribution: Found in the same areas as mesic dolomite prairie.

Dominant species: *little bluestem, bluejoint grass, sedge, tufted hair-grass, Indian grass, *cord grass.

Characteristic species: Ohio goldenrod, riddell goldenrod.

WET DOLOMITE PRAIRIE

Description: Prairie found on shallow, frequently saturated, soil over bedrock.

Distribution: Rare; this soil usually supports sedge meadow instead.

Dominant species: sedge, tufted hair-grass, *cord grass.

Characteristic species: prairie Indian-plantain.

HILL PRAIRIE

LOESS HILL PRAIRIE

Description: Prairie developed on deep loess on steep, exposed, south- or west-facing slopes.
Distribution: Mississippi, Illinois and Sangamon river bluffs.
Dominant species: *little bluestem, sideoats grama, Indian grass.
Characteristic species: green milkweed, false boneset, wild flax, *yellow puccoon, pale beard-tongue, *scurf-pea, *blue-eyed grass, *Spiranthes magnicamporum* (ladies' tresses).
Potential food plants: (Table 11:28); based on Evers (1955); see also the plants listed for limestone ledges.

GLACIAL DRIFT HILL PRAIRIE

Description: Glacial drift hill prairies are rather transient communities found on morainal ridges and steep slopes in east-central Illinois.
Distribution: Found along major rivers in the Grand Prairie section.
Dominant species: *little bluestem, sideoats grama, Indian grass.
Potential food plants: (Table 11:29); based on a study of five glacial drift prairies by Ebinger (1981).

GRAVEL HILL PRAIRIE

Description: Similar to gravel prairies but occur as openings in the forest; also similar to loess and glacial drift hill prairies.
Distribution: Found in northern Illinois and along major river valleys on coarse-grained glacial outwash.
Dominant species: *little bluestem, sideoats grama.
Characteristic species: frostweed, *prairie avens.
Potential food plants: (Table 11:30); based on Fell and Fell (1956).

SAND HILL PRAIRIE

Distribution: Prairie found on sand dunes atop river bluffs.
Dominant species: *little bluestem, sideoats grama, grama, *june grass.
Characteristic species: narrow-leaved goosefoot, *Lechea mucronata* (pinweed), *horsemint, salt-and-pepper plant, rock spikemoss, *goat's rue.

SHRUB PRAIRIE

SHRUB PRAIRIE

Description: Intergrades with mesic and wet-mesic sand prairie.
Distribution: Found in northern Illinois, the Kankakee Sand Area and the Chicago Lake Plain.
Dominant species: *big bluestem, *little bluestem, *black huckleberry, *swampy dewberry, *bristly blackberry, Indian grass, *hardhack, *low-bush blueberry.
Characteristic species: black chokeberry, purple chokeberry, tubercled orchid, lance-leaved violet, primrose violet.

SAVANNA

SAVANNA

Savannas are composed of an upper story of trees with 10 to 80% canopy cover and a nearly continuous ground cover of herbaceous species (Madany 1981). Savannas are transitional between forest and prairie and in presettlement Illinois they occurred either as broad ecotones separating the two or as isolated communities with one or the other. The location and precise nature of a savanna was determined by a complex interaction between topography and fire (Madany 1981). Three major subclasses are recognized based on soil differences. Savanna proper is the most typical and abundant, occurring on fine-textured soils derived from loess or glacial drift. Sand savanna occurs on sand deposits and is analogous to sand prairie. Barrens have no direct analog among prairie communities and have a strong forest element.

The term "barrens" frequently appears in early accounts of vegetation throughout the Midwest. Unfortunately, detailed studies of the botany of these areas were never made and the barrens themselves have almost entirely disappeared except for those in extreme environments. Those barrens remaining today are on steep slopes and narrow ridges in areas of rugged topography (Madany 1981).

DRY-MESIC SAVANNA

Description: Prairie composition similar to dry-mesic prairie.
Distribution: Throughout the prairie regions of Illinois.
Dominant species: *white oak, *bur oak, *post oak, *black oak, *little bluestem, Indian grass, *cord grass.
Characteristic species: *hazelnut, *American feverfew, carrion flower, starry campion.
Potential food plants: (Table 11:31-33); based on studies by Gleason (1910) and Madany (1981).

MESIC SAVANNA

Description: Soil moisture level and herbaceous vegetation similar to mesic prairie.
Distribution: Found primarily at the base of morainal ridges.
Dominant species: *white oak, *bur oak, *big bluestem, *little bluestem, Indian grass.
Characteristic species: *false sunflower, *veiny pea, golden alexanders.

SAND SAVANNA

DRY SAND SAVANNA

Description: Savanna characterized by very sandy soil with little humus and no A horizon; herbaceous plants less than 1 meter in height and with a low species diversity.
Distribution: Associated with dune and swale topography; found on dune crests in sand regions.
Dominant species: *little bluestem, sand reed, *Carex* spp. (sedge), *june grass, *black oak, *cord grass.
Characteristic species: *Commelina erecta* (day flower), *horsemint, cleft phlox
Potential food plants: (Table 11:32-34); based on Gleason (1910) and Madany (1981).

DRY-MESIC SAND SAVANNA

Description: Some soil A horizon developed.
Distribution: Found in same areas as dry sand savanna on lower or north- or east-facing slopes.
Dominant species: *little bluestem, *Carex* spp. (sedge), *black oak, Indian grass, *cord grass.
Characteristic species: flax-leaved aster, *New Jersey tea, clammy false foxglove, wild lupine, *prairie willow, low-bush blueberry.
Potential food plants: (Table 11:35-36); based on Gleason (1910).

BARRENS

Barrens were prairie expanses with scattered or widely spaced trees. In some situations, barrens apparently were prairie outliers which had been cut off from the larger prairies and which were being invaded by young trees and shrubs such as sumac and hazelnut (Vestal 1936). In other situations, the open aspect seems to have been preserved by fires periodically destroying the forest understory and young trees, thus keeping the canopy open (Curtis

1959:327). In either case, time and the cessation of prairie fires resulted in the disappearance of the barrens.

DRY BARREN

Description: Shallow soil or on dry, exposed slopes; tree layer composed of stunted xerophytic oaks, sparse grass cover less than 1 meter in height.

Distribution: Primarily found in the Shawnee Hills.

Dominant species: *little bluestem, *Carex pensylvanica* (sedge), poverty oat grass, *junegrass *blackjack oak, *post oak, *black oak, farkleberry.

Characteristic species: *black hickory, butterfly pea, *Liatris squarrosa* (blazing-star).

Potential food plants: (Table 11:37); based on Madany (1981).

DRY-MESIC BARREN

Description: Greater soil depth and moisture results in taller trees and grass than in the dry barren.

Distribution: Found in the Shawnee Hills and widely separate areas along major river valleys and at ravine crests.

Dominant species: *little bluestem, poverty oat grass, *white oak, *spanish oak, *black oak, Indian grass.

MESIC BARREN

Description: Greater soil depth or moisture resulting in still better growth of trees and grass.

Distribution: Found in the Cretaceous Hills; rare because this soil usually supports forest.

Dominant species: *big bluestem, *little bluestem, *white oak, *spanish oak, Indian grass.

WETLAND

Plant communities of aquatic environments are relatively homogeneous throughout Illinois since the primary factors determining the species composition are water depth and duration of flooding. The type of substrate, whether the bottom is rocky, sandy, or muddy, is important for the growth of some, but not all, plants as well. Current speed is important in that relatively few plants can become established in the fast-moving water of rivers or streams. The majority of aquatic plants are limited to ponded water or to ponds, slow-moving streams, or backwaters. Certain plants with northern affinities occur only in northern Illinois and in the cooler water of springs and spring branches of southern Illinois. Water pH, determined by substrate material and water source, also influences the type of plants growing in a particular aquatic situation. The vegetation of northern bogs, such as those of northeastern Illinois which occupy old glacial depressions, are dominated by acidic mosses and characteristic acidophilous plants. Alkaline marshes known as fens support a very different flora dominated by sedges. Fens have alkaline water sources such as calcareous springs or seeps or occur in low ground in areas with calcareous bedrock. The glacial lakes of the Northeastern Morainal Division can have either a peat, sand, or marl base, each with a different biota. With time, peat lakes generally fill in to produce bogs, marl lakes to form fens.

MARSH

MARSH

Description: Marsh communities are dominated by tall graminoid plants and occupied by standing water for most of the year.

Distribution: Freshwater marshes are common in glacial potholes, river valleys, and on lake plains. Although originally widespread, many have been drained.

Dominant species: *Carex lacustris* (sedge), *Decodon verticillatus* (swamp loosestrife), *reed grass, **Polygonum amphibium* (water smartweed), **P. coccineum* (water smartweed),

Scirpus fluviatilis (bulrush), *soft-stem bulrush, *narrow-leaved cattail, *broad-leaved cattail.
Characteristic species: water-plantain, false aster, mermaid-weed, *common arrowhead, *marsh skullcap.
Potential food plants: Table 12:38-39; Based on Solomon (1979), Sampson (1921), and A. Jones and Bell (1974).

BRACKISH MARSH
Description: Marshes found in salty, seepy areas.
Distribution: Rare; found along a short segment of river bluff upstream from Starved Rock.
Dominant species: *cord grass.
Characteristic species: spear scale, swamp rose mallow, bayonet-grass.

SWAMP

Swamps are wetlands dominated by woody plants. They are developed on heavy soils in the broad, flat and frequently flooded bottomlands of the Mississippi, Cache, and Ohio rivers. The relatively warm climate of southern Illinois allows the presence of species with more southeasterly affinities, particularly in the Coastal Plain Section. The vegetation is dominated by such species as bald cypress, water tupelo, and water locust (Schwegman *et al.* 1973).

SWAMP
Description: Forested wetlands on permanent or semipermanent bodies of water.
Distribution: Extreme southern Illinois.
Dominant species: tupelo gum, bald-cypress, *buttonbush.
Characteristic species: *black willow, pumpkin ash, swampy rose, Virginia willow.
Potential food plants: Table 12:40-41; based on Mohlenbrock *et al.* (1961), Mohlenbrock (1959), and Anderson and White (1970).

SHRUB SWAMP
Description: Wetland characterized by at least 50% coverage by shrubs (less coverage is termed a pond) and less than 20% coverage by trees (greater coverage is termed a swamp).
Distribution: Often associated with ponds in wet floodplain forests or in glacial potholes with tall shrub bog communities; throughout Illinois.
Dominant species: *buttonbush, *red-osier dogwood, *pussy willow, *sandbar willow, *speckled alder.

BOG

Wetland characterized by low nutrient, acid peat deposits, usually in glacial depressions with restricted drainage. They contain distinctive plants such as pitcher-plant, sundew, leatherleaf, poison sumac, winterberry, and dwarf birch.

GRAMINOID BOG
Description: Floating community closest to open water; composed primarily of sedges and *Sphagnum* moss.
Distribution: Northeastern Morainal Division.
Dominant species: *Carex* spp. (including *C. hystricina*, *C. lasiocarpa*, *C. haydenii* and others), *Polytrichum* and *Sphagnum* mosses.
Characteristic species: *sundew, buckbean, *pitcher-plant.
Potential food plants: Table 12:42; compiled from Sheviak and Haney (1973), includes all types of bogs.

TABLE 12. Potential Food Plants of Aquatic Communities

SCIENTIFIC NAME	COMMON NAME	USE*	Western Ill. Marsh 38**	Central Ill. Marsh 39	Southern Ill. Swamp 40	Southern Ill. Cypress Swamp 41	Bogs 42	Fens 43	Sedge Meadow 44	Seeps and Springs 45	Eastern Ill. Hillside Marsh 46	Floodplain Ponds 47	Southern Ill. Sinkholes 48	Oxbow Lake 49	Prairie Potholes 50	Ponds 51	Sangamon Drainage 52	Southern Wisc. Lake 53	Southern Ill. Lake 54	Eastern Ill. Lake 55	TOTAL
ACER NEGUNDO	Boxelder	C2/B2					*														1
A. SACCHARINUM	Silver Maple	C2/B2					*														1
ACORUS CALAMUS	Sweet Flag	T3	*	*	*			*				*							*	*	7
ALLIUM CANADENSE	Wild Garlic	T28								*											1
A. STELLATUM	Cliff Onion	T28													*						1
AMPHICARPA BRACTEATA	Hog Peanut	F28									*				*						2
APIOS AMERICANA	Groundnut	T9									*	*									2
ARUNDINARIA GIGANTEA	Giant Cane	S4			*																1
ASCLEPIAS SYRIACA	Common Milkweed	G345													*						1
A. TUBEROSA	Butterfly-weed	G345/T67													*						1
ASTRAGALUS CANADENSIS	Canadian Milk Vetch	T37													*						1
BETULA LUTEA	Yellow Birch	B2					*														1
CALTA PALUSTRIS	Marsh Marigold	G3	*				*		*		*										4
CLAYTONIA VIRGINICA	Spring Beauty	T3									*										1
CORYLUS AMERICANA	Hazelnut	N6					*														1
CORNUS STOLONIFERA	Red Osier	F6					*	*													2
CYPERUS ARISTATUS	Plantain-leaved Sedge	T28	*							*											2
C. ESCULENTUS	Nut-grass	T28		*	*	*													*		4
ELYMUS CANADENSIS	Nodding Wild Rye	S7													*						1
FRAGARIA VIRGINIANA	Wild Strawberry	F4							*						*						2
FRAXINUS PENNSYLVANICA	Green Ash	CE			*																1
GLYCERIA BOREALIS	Northern Manna Grass	S56																*			1
G. STRIATA	Fowl Manna Grass	S56	*		*		*	*	*		*	*							*	*	9
HELIANTHUS ANNUUS	Sunflower	S67									*										1
HERACLEUM MAXIMUM	Cow Parsnip	T4/G4	*	*																	2
HORDEUM JUBATUM	Squirrel-tail Grass	S56							*						*						2
KOELERIA CRISTATA	Junegrass	S5													*						1
LATHYRUS PALUSTRIS	Marsh Vetchling	S6						*	*						*						3
LYCOPUS ASPER	Rough Water Horehound	T28		*																	1
L. UNIFLORUS	Northern Bugle Weed	T28		*				*	*												3

TABLE 12. (Continued).

SCIENTIFIC NAME	COMMON NAME	USE*	Western Ill. Marsh 38	Central Ill. Marsh 39	Southern Ill. Swamp 40	Southern Ill. Cypress Swamp 41	Bogs 42	Fens 43	Sedge Meadow 44	Seeps and Springs 45	Eastern Ill. Hillside Marsh 46	Floodplain Ponds 47	Southern Ill. Sinkholes 48	Oxbow Lake 49	Prairie Potholes 50	Ponds 51	Sangamon Drainage 52	Southern Wisc. Lake 53	Southern Ill. Lake 54	Eastern Ill. Lake 55	TOTAL
MAIANTHEMUM CANADENSE	Wild Lily-of-the-Valley	F678					*														1
NELUMBO LUTEA	American Lotus	T28/S67/G34			*							*			*	*	*		*		6
NUPHAR ADVENA	Yellow Pond Lily	T28			*			*								*	*		*		5
NYMPHAEA ODORATA	Fragrant Water Lily	G3			*			*							*		*				4
N. TUBEROSA	White Water Lily	T28/S67/G3	*									*			*	*	*				5
ONOCLEA SENSIBILIS	Sensitive Fern	T8		*			*	*						*							4
OSMUNDA CINNAMOMEA	Cinnamon Fern	G3		*			*														2
OXALIS STRICTA	Yellow Wood Sorrel	G4-7													*						1
O. VIOLACEA	Violet Wood Sorrel	G4-7/T													*						1
PANICUM CAPILLARE	Witch Grass	S67							*												1
PARTHENOCISSUS QUINQUEFOLIA	Virginia Creeper	F7				*															1
PEDICULARIS CANADENSIS	Lousewort	G3						*							*						2
P. LANCEOLATA	Swamp Wood Betony	G3		*							*										2
PELTANDRA VIRGINICA	Arrow Arum	T28		*	*														*		3
PETALOSTEMUM CANDIDUM	White Prairie Clover	T?													*						1
P. PURPUREUM	Purple Prairie Clover	T?													*						1
PHRAGMITES COMMUNIS	Reed Grass	G2/T1/S5	*	*			*	*										*			5
PHYSALIS HETEROPHYLLA	Ground Cherry	F67													*					*	2
P. VIRGINIANA	Ground Cherry	F67													*						1
PODOPHYLLUM PELTATUM	Mayapple	F6									*										1
POLYGONUM COCCINEUM	Water Smartweed	S56	*	*	*			*	*						*						6
POPULUS DELTOIDES	Cottonwood	CE			*														*		2
P. TREMULOIDES	Trembling Aspen	CE					*														1
PRUNUS AMERICANA	American Plum	F6													*						1
P. PENNSYLVANICA	Pin Cherry	F56		*																	1
P. SEROTINA	Black Cherry	F56													*						1
QUERCUS LYRATA	Overcup Oak	N7			*																1
Q. PHELLOS	Willow Oak	N7			*																1
RIBES AMERICANA	Wild Black Currant	F5													*						1

TABLE 12. (Concluded).

SCIENTIFIC NAME	COMMON NAME	USE	Western Ill. Marsh 38	Central Ill. Marsh 39	Southern Ill. Swamp 40	Southern Ill. Cypress Swamp 41	Bogs 42	Fens 43	Sedge Meadow 44	Seeps and Springs 45	Eastern Ill. Hillside Marsh 46	Floodplain Ponds 47	Southern Ill. Sinkholes 48	Oxbow Lake 49	Prairie Potholes 50	Ponds 51	Sangamon Drainage 52	Southern Wisc. Lake 53	Southern Ill. Lake 54	Eastern Ill. Lake 55	TOTAL
RORIPPA ISLANDICA	Marsh Yellow Cress	G3		*					*			*		*							4
SAGITTARIA LATIFOLIA	Common Arrowhead	T6-8	*		*		*	*		*	*	*	*	*			*	*	*		12
SAMBUCUS CANADENSIS	Elderberry	F67					*	*							*						3
SCIRPUS ACUTUS	Hard-stem Bulrush	T28/G3/S67	*		*												*	*	*		5
S. VALIDUS	Soft-stem Bulrush	T28/P5/S4-7	*					*	*	*	*		*				*			*	8
SMILACINA STELLATA	False Solomon's Seal	F5						*													1
SPARGANIUM EURYCARPUM	Bur-reed	T28	*	*	*		*	*						*				*			7
STAPHYLEA TRIFOLIA	Bladdernut	S67											*								1
SYMPHORICARPOS OCCIDENTALIS	Wolfberry	F67		*																	1
TYPHA ANGUSTIFOLIA	Narrow-leaved Cattail	T128/G34/P4	*	*														*	*		4
T. LATIFOLIA	Common Cattail	T128/G34/PS4	*	*			*	*	*		*	*	*	*				*	*	*	12
URTICA DIOICA	Stinging Nettle	G345					*														1
VACCINIUM CORYMBOSUM	High-bush Blueberry	F6					*														1
V. MACROCARPON	American Cranberry	F6					*														1
VERBENA HASTATA	Blue Vervain	S6		*					*						*					*	4
VIBURNUM LENTAGO	Nannyberry	F67		*																	1
VITIS RIPARIA	Frost Grape	F6										*	*								2
V. VULPINA	Frost Grape	F7							*				*								2
XANTHIUM COMMUNE	Cocklebur	S7							*			*									2
ZIZANIA AQUATICA	Wild Rice	S6		*								*						*	*		4
TOTAL			14	20	16	2	19	17	14	4	11	11	6	5	26	3	7	8	12	6	201

* C=Bark or Cambium, F=Fruit, G=Greens, S=Seeds, N=Nuts, T=Roots or tubers, P=Pollen, 1=Winter, 2=Early Spring, 3=Spring, 4=Early summer, 5=Late Summer, 6=Early Fall, 7=Fall, 8=Late Fall, 9=All Year, E=Emergency

** See Table 3 For References.

LOW SHRUB BOG

Description: May or may not be floating; comprised of low shrubs over a moss stratum.
Distribution: Northeastern Morainal Division.
Dominant species: *dwarf birch, *leatherleaf, *Polytrichum* and *Sphagnum* mosses.
Characteristic species: cotton grass, *American cranberry.
Potential food plants: Table 11:42; compiled from Sheviak and Haney (1973);includes all types of bog.

FORESTED BOG

Description: Fairly well consolidated peat supports a tree layer with greater than 20% canopy cover and tall shrubs.
Distribution: Northeastern Morainal Division.
Dominant species: winterberry, *American larch, poison sumac.
Characteristic species: *Lady's slipper orchid, shining clubmoss, *cinnamon fern, *star-flower, *high-bush blueberry.
Potential food plants: Table 11:42; compiled from Sheviak and Haney (1973); includes all types of bog.

TALL SHRUB BOG

Description: Considered climax in Illinois bog succession; lacking the tree layer.
Distribution: Northeastern Morainal Division.
Dominant species: *red osier dogwood, winterberry, poison sumac.
Potential food plants: Table 11:42 compiled from Sheviak and Haney (1973); includes all types of bog.

FEN

Wetland often associated with strongly calcareous springs, seeps, sedge meadows, and marshes. The soil is an organic peat that is neutral or alkaline in reaction compared to bogs which have acid peat and marshes which are developed on mineral soil (Curtis 1959). They are usually associated with swales in areas of calcareous groundwater, filled glacial lakes, or calcareous springs and seeps.

CALCAREOUS FLOATING MAT

Description: Floating layer of sedge peat dominated by sedges and grasses.
Distribution: Found in the northern one-third of Illinois, extending down the Illinois River.
Dominant species: bluejoint grass, *Carex* sp. (sedge), *Decodon verticillatus* (swamp loosestrife).
Characteristic species: buckbean, marsh cinquefoil, hoary willow, *Salix pedicellaris* (willow).
Potential food plants: Table 11:43; compiled from Curtis's(1959) study in southern Wisconsin, Sherff's (1913) study of Skokie Swamp in northeastern Illinois, and Moran's (1981) study of 12 fens in northeastern Illinois; includes all types of fen.

GRAMINOID FEN

Description: Developed on sloping peat at the edge of a moraine or as a raised island in a marsh or sedge meadow.
Distribution: Occurs in the northern one-third of the state, extending down the Illinois River.
Dominant species: *big bluestem, *little bluestem, *Carex* sp. (sedge), indian grass, prairie dropseed.
Characteristic species: *Carex* spp. (sedge), marsh blazing-star, Kalm's lobelia, *Lysimachia quadriflora* (loosestrife), *Muhlenbergia glomerata* (muhly), grass-of-parnassus, Ohio goldenrod.
Potential food plants: See calcareous floating mat.

LOW SHRUB FEN

Description: Similar to calcareous floating mat with the addition of low shrubs.
Distribution: Occurs in the northern one-third of Illinois, extending down the Illinois River.

Dominant species: *Carex* sp.(sedge), shrubby cinquefoil.
Potential food plants: See calcareous floating mat.

TALL SHRUB FEN

Description: Similar to a low shrub fen with the addition of tall shrubs.
Distribution: Found on the edge of upper Lake Peoria.
Dominant species: *red osier dogwood, poison sumac, pussy willow.
Characteristic species: *showy lady's slipper, *queen-of-the-prairie, spreading goldenrod.
Potential food plants: See calcareous floating mat.

FORESTED FEN

Description: Relatively steep slopes in peat; tree coverage greater than 20%.
Distribution: Occurs in the northern one-third of Illinois, extending down the Illinois River.
Dominant species: *black ash, *American larch, *arbor-vitae.
Characteristic species: hemlock parsley, *purple avens, green orchid, *skunk cabbage.
Potential food plants: See calcareous floating mat.

SEDGE MEADOW

Description: Sedge meadow, dominated by sedge and water-tolerant shrubs, is a stage between marshes dominated by cattails and bulrushes and wet prairie dominated by grass.
Distribution: Sedge meadow originally occurred rather commonly in the Northeastern Morainal Division (Sherff 1913, Curtis 1959:580-1) on peat, muck or wet sand.
Dominant species: *Carex* spp. (sedge), bluejoint grass.
Characteristic species: shiny swamp aster, *white turtlehead, bog willow herb, *spotted joe-pye weed, marsh St. John's-wort.
Potential food plants: Table (12:44); based on Sherff (1913) and Curtis (1959, 580-81).

PANNE

PANNE

Description: Similar to graminoid fen and calcareous seep.
Distribution: Wet and wet-mesic swales in calcareous sand within 1 mile of Lake Michigan.
Dominant species: bluejoint grass, twig rush, *Carex* spp. (sedge), *Juncus balticus* var. *littoralis* (rush), shrubby cinquefoil.
Characteristic species: *Carex* spp. (sedge), *Eleocharis olivacea* (spike rush), *Linum medium* var. *texanum* (wild flax), *Triglochin maritima* and *T. palustris* (arrow-grass), horned bladderwort.

SEEP & SPRING

Seeps and springs occur along the base of outcrops where water has accumulated because of fractures or by following bedding planes.

SEEP

Description: Area with saturated soil caused by diffuse flow of water to the surface.
Distribution: Throughout Illinois.
Dominant species: *Carex* spp.(sedge), *alternate-leaved dogwood, black ash, *skunk cabbage, *fowl manna grass, *spotted touch-me-not, clearweed.
Characteristic species: *swamp wood betony, *white turtlehead, spreading goldenrod, cinnamon willow herb.

Potential food plants: (Table 12:45–46); based on Voigt and Mohlenbrock 1964 and also studies of Hillside marsh in east-central Illinois by Phipps and Speer (1958) and Phillippe and Ebinger (1977).

ACID GRAVEL SEEP

Description: Seep characterized by muck or peat deposits and low pH.

Distribution: Cretaceous Hills Section.

Characteristic species: *lady fern, *Carex incomperta* (sedge), *cinnamon fern, *royal fern, *Sphagnum* moss, screw-stem, netted chain fern.

CALCAREOUS SEEP

Description: Seep so calcareous that tufa deposits form.

Distribution: Wisconsin till plain.

Characteristic species: twig rush, tufted hair-grass, *Eleocharis rostella* (spike rush), *Juncus brachycephalus* (rush), shrubby cinquefoil, *Rhyncospora capillacea* and *R. alba* (beaked rush), *Scirpus caespitosus* (bulrush), *Scleria verticillata* (nut rush), prairie-dock, false asphodel, *Triglochin palustris* (arrow-grass).

SAND SEEP

Description: Usually acid seepage through sand.

Distribution: Best developed in the Chicago Lake Plain and the Kankakee Sand area.

Characteristic species: *lady fern, *fowl manna grass, *cinnamon fern, *royal fern, *ninebark, *skunk cabbage, spinulose woodfern.

SPRING

Description: Concentrated flow of water from a definite orifice.

Distribution: Ozark Hills and Shawnee Hills Division.

Characteristic species: *Chara* spp.

POND & LAKE

Bodies of open, standing water lacking emergent woody or graminoid vegetation.

POND

A small, still body of water, usually shallow enough to allow rooted aquatic vegetation.

POND

Distribution: Almost all bodies of water in Illinois, including many backwater sloughs connected to major rivers.

Characteristic species: *pond lily, *white waterlily, pondweed, *Spirodela* spp. and *Lemna* spp.(duckweed), *Polygonum* sp. (water smartweed).

Potential food plants: (Table 12:47–52) The potential food plants have been compiled from several types of ponds. These include a reservoir lake in east-central Illinois (Phillippe and Ebinger 1977), a series of reservoir lakes, ponds and sinkholes in southern Illinois (Mohlenbrock *et al.* 1961), ponds on the Mississippi River floodplain at Mississippi Palisades State Park in Carroll County (Wunderlin 1966), the Sangamon River basin (A. Jones and Bell 1974:3–4), sinkhole ponds in southern Illinois, an oxbow lake in the floodplain of Haw Creek, a tributary of the Spoon River, in Knox County (Weik and Baker 1975), a study of prairie potholes (Glenn-Lewin and Crist 1981), and 197 ponds in 14 counties in east-central Illinois (Vogel and Ebinger 1979).

LAKE

A lake has at least a section of barren, wave-swept shoreline without attached aquatic plants.

LAKE

Distribution: Formerly on upland plains and bottomlands throughout Illinois; most have been drained. Examples remain in the Northeastern Morainal Division and a few backwater lakes still present along the Illinois and Mississippi rivers.

Potential food plants: (Table 12:53–55); based on Curtis' (1959:584–586) list of plants found in glacial lakes, Phillippe and Ebinger's (1977) study of a reservoir lake in east-central Illinois, and Mohlenbrock *et al.* (1961) in their study of southern Illinois lakes and ponds. No potential food plants occur among the submerged species; however, a number do occur among the emergent aquatic plants; for this reason the potential food plants of lakes and ponds are very similar.

GREAT LAKE

Description: Lake Michigan.

STREAM

Permanent flowing water.

CREEK

Perennial stream with a watershed smaller than 200 square miles (520 square kilometers). Intermittent streams are classified with the communities in which they occur.

LOW-GRADIENT CREEK

Description: Gradient less than 1 foot/mile; sluggish current, no riffles; sediment composed of silt and organic matter.

Distribution: Prairie uplands and the bottomlands of major rivers.

Potential food plants: Shallow, slow-moving water along margins--American lotus, calamus, arrow-arum, arrowhead, bulrush, common cattail, water smartweed; deep, slow-moving water--yellow pond lily, white water lily (Winterringer and Lopinot 1966).

MEDIUM-GRADIENT CREEK

Description: Gradient 1–10 feet/mile.

Distribution: Throughout Illinois.

Potential food plants: See low-gradient stream.

HIGH-GRADIENT CREEK

Description: Gradient 10 feet/mile or greater; riffles, pools, and sand and gravel bars characteristic.

Distribution: Throughout Illinois, many headwater streams are examples.

Potential food plants: See low-gradient stream.

RIVER

Stream with a watershed of 200 square miles or more.

LOW-GRADIENT RIVER

Description: Gradient less than 1 foot/mile; meandering with a sluggish current; sediment mostly silt.

Distribution: Throughout Illinois, eg., Big Muddy, Sangamon, Green rivers.

Potential food plants: See low-gradient stream.

MEDIUM-GRADIENT RIVER

Description: Gradient 1–10 feet/mile; gravel riffles and raceways and sandbars characteristic.

Distribution: Throughout Illinois, eg., sections of the Rock, Kankakee and Mackinaw rivers.

Potential food plants: See low-gradient stream.

MAJOR RIVER
Description: Very wide and deep channel and very large flow.
Distribution: Mississippi, Illinois, Ohio, and Wabash rivers.
Potential food plants: See low gradient stream.

PRIMARY

Natural communities found where soil is thin or absent and maintained at an early successional stage by substrate or natural disturbance.

GLADE

Opening in forest caused by thin soil and/or steep south or west exposures. Usually a mosaic of stunted trees, shrubs, and herbaceous vegetation and bare patches.

SANDSTONE GLADE
Description: Sparse, stunted trees on poorly developed soil on sandstone outcrops.
Distribution: Found on the tops of sandstone cliffs and steep upper slopes of south-facing escarpments in the Shawnee Hills.
Dominant species: *little bluestem, *red cedar, *blackjack oak, *post oak, winged elm, farkleberry.
Characteristic species: hairy lip fern, doveweed, tufted hair-grass, slender false foxglove, pineweed, widow's cross, poverty grass, flower-of-an-hour, *prickly-pear.
Potential food plants: (Table 13:56); based on Winterringer and Vestal (1956).

LIMESTONE GLADE
Description: Overlaps with hill prairie although there are relatively more trees, shrubs and vines; found on steep south- and west-facing spurs and limestone bluffs with shallow, rocky, often clayey soil.
Distribution: Shawnee Hills Division and on Mississippi and Illinois river bluffs.
Dominant species: *little bluestem, sideoats grama, indian grass.
Characteristic species: American aloe, spreading aster, prairie indian-plantain, *rattlesnake master, *red cedar, false boneset, hoary puccoon, false dragonhead, *chinquapin oak, prairie-dock, catbrier, pale coneflower.
Potential food plants: (Table 13:57); based on Ozment (1967), Evers (1955), and Kurz (1981).

SHALE GLADE
Description: Natural openings on shale outcrops.
Distribution: Southern section of the Ozark Division.
Dominant species: *blackjack oak, *post oak, *red cedar, *little bluestem.
Characteristic species: butterfly weed.

CLIFF

Vertical exposure, generally lacking soil.

SANDSTONE CLIFF
Distribution: Shawnee Hills and scattered elsewhere.
Characteristic species: marginal fern, late alumroot, hairy lip fern.

LIMESTONE CLIFF
Distribution: Throughout Illinois, most prominent along the Illinois and Mississippi Rivers.
Characteristic species: bladder fern, purple cliffbrake fern, smooth cliffbrake fern.

DOLOMITE

Description: Very similar to limestone cliff.
Distribution: Stream valleys in the northern part of the state.
Characteristic species: bladder fern, *ninebark, *American spikenard, *bellflower, smooth cliffbrake fern.

SANDSTONE OVERHANG

Description: Sheltered, shaded overhangs.
Distribution: Shawnee Hills division.
Characteristic species: filmy fern, French's shooting-star.

ERODING BLUFF

Description: Vertical exposures of eroded, unconsolidated material or weak rock.
Distribution: Common along major rivers in the glaciated part of the state.
Characteristic species: tufted hair-grass, *field goldenrod, yellow pimpernel.

LAKE SHORE

Shoreline communities along Lake Michigan formed on lake-deposited sands.

BEACH

Description: Recently deposited sands which are nearly bare to grass-covered.
Dominant species: beach grass, sand reed, *nodding wild rye.
Characteristic species: sea rocket, common bugseed.

FOREDUNE

Description: Fairly dense cover of low shrubs and grass similar to dry sand prairie.
Dominant species: *little bluestem, bearberry, trailing juniper.
Characteristic species: *common juniper, *purple prairie clover.
Potential food plants: (Table 13:58); based on Gates (1912).

CULTURAL

DISTURBANCE

Description: While this community is not listed by White and Madany (1978) who delineate specific types of cultural influence, disturbance of the landscape is extremely common at present and would certainly have also resulted from prehistoric human activities.
Distribution: Throughout Illinois.
Dominant species: Native and introduced ruderal plants, especially annuals and biennials with high seed production and extremely efficient dispersal mechanisms, or perennials with aggressive vegetative reproduction (A. Jones and Bell 1974:9).
Potential food plants: (Table 13:59–63) Lists are included for central Illinois (A. Jones and Bell 1974), eastern Illinois (Phillippe and Ebinger 1977), southern Illinois (Mohlenbrock 1967, Heineke 1978), southern Wisconsin (Curtis 1959:591–2]), and middle Mississippi River sand and mud flats (Evans 1979).

TABLE 13. Potential Food Plants of Disturbance Communities

SCIENTIFIC NAME	COMMON NAME	USE*	Sandstone Ledge 56**	Limestone Glade 57	Upper Beach Lake Michigan 58	Central Ill. Disturbance 59	Southern Ill. Disturbance 60	Eastern Ill. Disturbance 61	Central Miss. Sand & Mudflats 62	Southern Wisc. Disturbance 63	TOTAL
ACER NEGUNDO	Boxelder	C2,B2		*			*	*	*		4
A. RUBRUM	Red Maple	C2,B2		*							1
A. SACCHARINUM	Silver Maple	C2,B2		*			*		*		3
ALLIUM CANADENSE	Wild Garlic	T28		*	*		*	*			4
A. CERNUUM	Wild Onion	T28			*						1
A. STELLATUM	Cliff Onion	T28		*							1
AMARANTHUS HYBRIDUS	Green Amaranth	G3,S56					*	*			2
A. RETROFLEXUS	Rough Pigweed	G3,S56				*	*		*	*	4
AMELANCHIER ARBOREA	Shadbush	F4	*	*							2
AMPHICARPA BRACTEATA	Hog Peanut	F28		*							1
APIOS AMERICANA	Groundnut	T9					*				1
ARISAEMA TRIPHYLLUM	Jack-in-the-Pulpit	T28		*							1
ARUNDINARIA GIGANTEA	Giant Cane	S5		*							1
ASCLEPIAS SYRIACA	Common Milkweed	G345			*	*	*	*		*	5
A. TUBEROSA	Butterfly-weed	G345,T67		*	*		*				3
ASIMINA TRILOBA	Pawpaw	F7		*							1
ASTRAGALUS CANADENSIS	Canadian Milk Vetch	T37			*						1
CARYA GLABRA	Pignut Hickory	N7	*	*			*				3
C. LACINIOSA	Shellbark Hickory	N7		*							1
C. OVALIS	Sweet Pignut Hickory	N7		*							1
C. OVATA	Shagbark Hickory	N7		*							1
C. TEXANA	Black Hickory	N7		*							1
C. TOMENTOSA	Mockernut Hickory	N7		*							1
CELASTRUS SCANDENS	Bittersweet	CE		*				*			2
CELTIS OCCIDENTALIS	Hackberry	F67		*			*				2
CLAYTONIA VIRGINICA	Spring Beauty	T3		*							1
CORYLUS AMERICANA	Hazelnut	N6		*			*				2
CRATAEGUS CALPODENDRON	Hawthorn	F67		*							1
C. MOLLIS	Red Haw	F67						*			1
C. PRUINOSA	Hawthorn	F67		*							1
CYPERUS ARISTATUS	Plantain-leaved Sedge	T28							*		1
C. ESCULENTUS	Nut-grass	T28					*				1
DENTARIA LACINIATA	Toothwort	T23		*							1
DIOSPYROS VIRGINIANA	Persimmon	F7	*	*			*	*	*		3
ELYMUS CANADENSIS	Nodding Wild Rye	S7		*	*		*	*	*		3
FAGUS GRANDIFOLIA	Beech	N7		*							1
FRAGARIA AMERICANA	Hillside Strawberry	F4					*				1
F. VIRGINIANA	Wild Strawberry	F4			*	*	*	*	*		4
GLYCERIA STRIATA	Fowl Manna Grass	S56						*			1
GYMNOCLADUS DIOICA	Kentucky Coffee Tree	S7		*							1
HELIANTHUS ANNUUS	Sunflower	S67				*					1
H. TUBEROSUS	Jerusalem Artichoke	T278					*	*			2
HERACLEUM MAXIMUM	Cow Parsnip	T4/G4			*						1
HORDEUM PUSILLUM	Little wild barley	S4				*	*	*			3
HYDROPHYLLUM APPENDICULATUM	Great Waterleaf	G3		*							

TABLE 13. (Continued).

SCIENTIFIC NAME	COMMON NAME	USE	Sandstone Ledge 56	Limestone Glade 57	Upper Beach Lake Michigan 58	Central Ill. Disturbance 59	Southern Ill. Disturbance 60	Eastern Ill. Disturbance 61	Central Miss. Sand & Mudflats 62	Southern Wisc. Disturbance 63	TOTAL
IPOMOEA PANDURATA	Wild Sweet Potato	TE		*		*	*	*			4
IVA ANNUA	Sumpweed	S56				*					1
JUGLANS NIGRA	Black Walnut	N7		*			*				2
KOELERIA CRISTATA	June Grass	S5		*							1
LATHYRUS PALUSTRIS	Marsh Vetchling	S6				*					1
MALUS CORONARIA	Wild Crabapple	F6					*				1
M. IOENSIS	Iowa Crabapple	F6		*				*			2
MORUS RUBRA	Red Mulberry	F5		*			*	*	*		4
OXALIS STRICTA	Yellow Wood Sorrel	G4-7		*		*	*		*		4
O. VIOLACEA	Violet Wood Sorrel	G3-7,T?	*	*							2
PANICUM CAPILLARE	Witch Grass	S67		*		*	*	*	*	*	6
PARTHENOCISSUS QUINQUEFOLIA	Virginia Creeper	F7		*		*	*				3
PEDICULARIS LANCEOLATA	Swamp Wood Betony	G3			*						1
PETALOSTEMUM CANDIDUM	White Prairie Clover	S?,T?		*	*						2
P. PURPUREUM	Purple Prairie Clover	S?,T?		*	*						2
PHYSALIS HETEROPHYLLA	Ground Cherry	F67		*		*	*			*	4
P. PUBESCENS	Annual Ground Cherry	F67		*		*	*		*		4
P. VIRGINIANA	Ground Cherry	F67		*							1
PHYTOLACCA AMERICANA	Pokeberry	G3		*		*	*	*			4
PODOPHYLLUM PELTATUM	Mayapple	F6		*							1
POLYGONUM COCCINEUM	Water Smartweed	S56					*		*		2
POLYGONUM ERECTUM	Knotweed	S56				*	*				2
POPULUS DELTOIDES	Cottonwood	CE		*			*	*	*		4
PRUNUS AMERICANA	American Plum	F6		*		*	*				3
P. ANGUSTIFOLIA	Chickasaw Plum	F56		*							1
P. SEROTINA	Black Cherry	F56		*		*	*	*			4
P. VIRGINIANA	Chokecherry	F67		*							1
PYCNANTHEMUM VIRGINIANUM	Mountain Mint	G3			*						1
QUERCUS ALBA	White Oak	N7		*							1
Q. IMBRICARIA	Shingle Oak	N7		*				*			2
Q. MARILANDICA	Blackjack Oak	N7	*	*							2
Q. MUHLENBERGII	Yellow Chestnut Oak	N7		*							1
Q. RUBRA	Red Oak	N7		*			*				2
Q. STELLATA	Post Oak	N7	*	*							2
Q. VELUTINA	Black Oak	N7		*							1
RHUS COPALLINA	Dwarf Sumac	F67	*	*			*				3
R. GLABRA	Smooth Sumac	F67		*				*			2
RIBES CYNOSBATI	Prickly Gooseberry	F6		*							1
RORIPPA ISLANDICA	Marsh Yellow Cress	G3							*		1
RUBUS ALLEGHENIENSIS	Common Blackberry	F6		*		*	*	*			4
R. ARGUTUS	High-bush Blackberry	F45					*	*			2
R. FLAGELLARIS	Dewberry	F56		*		*	*	*			4
R. OCCIDENTALIS	Black Raspberry	F56		*			*	*			3
R. TRIVALIS	Southern Dewberry	F56							*		1
RUDBECKIA LACINIATA	Goldenglow	G3					*				1

TABLE 13. (Concluded).

SCIENTIFIC NAME	COMMON NAME	USE*	Sandstone Ledge 56	Limestone Glade 57	Upper Beach Lake Michigan 58	Central Ill. Disturbance 59	Southern Ill. Disturbance 60	Eastern Ill. Disturbance 61	Central Miss. Sand & Mudflats 62	Southern Wisc. Disturbance 63	TOTAL
SAGITTARIA LATIFOLIA	Common Arrowhead	T6-8							*		1
SAMBUCUS CANADENSIS	Elderberry	F67					*	*	*		3
SMILACINA RACEMOSA	False Solomon's Seal	T?,F5		*							1
SMILAX BONA-NOX	Catbrier	T28	*	*				*			3
S. GLAUCA	Catbrier	T28	*	*			*		*		4
S. ROTUNDIFOLIA	Catbrier	T28		*							1
SOLANUM AMERICANUM	Black Nightshade	F67		*		*	*	*	*		5
SPOROBOLUS CRYPTANDRUS	Sand Dropseed	S67		*					*		2
STAPHYLEA TRIFOLIA	Bladdernut	S67		*							1
STROPHOSTYLES HELVOLA	Wild Bean	T28		*			*				2
TILIA AMERICANA	Basswood	B2		*							1
TYPHA LATIFOLIA	Common Cattail	T128/G34/PS4							*		1
ULMUS RUBRA	Slippery Elm	CE		*			*	*	*		4
URTICA DIOICA	Stinging Nettle	G345							*	*	2
VACCINIUM VACILLANS	Low-bush Blueberry	F6		*							1
VERBENA HASTATA	Blue Vervain	S6				*	*	*			3
VITIS AESTIVALIS	Summer Grape	F7		*				*			2
V. CINEREA	Winter Grape	F7		*			*				2
V. RIPARIA	Frost Grape	F6		*				*	*		3
V. VULPINA	Frost Grape	F7		*		*	*	*			4
VULPIA OCTOFLORA	Six-weeks Fescue	S?		*							1
XANTHIUM COMMUNE	Cocklebur	S?		*							1
TOTAL			9	81	12	22	48	33	24	5	234

* C=Bark or Cambium, F=Fruit, G=Greens, S=Seeds, N=Nuts, T=Roots or tubers, P=Pollen, 1=Winter, 2=Early Spring, 3=Spring, 4=Early summer, 5=Late Summer, 6=Early Fall, 7=Fall, 8=Late Fall, 9=All Year, E=Emergercy

** See Table 3 For References.

LITERATURE CITED

Adams, D. E. and R. C. Anderson
1980 Species Response to a Moisture Gradient in Central Illinois Forests. *American Journal of Botany*, Vol. 67, pp. 381-392.

Albertson, F. W. and J. E. Weaver
1945 Injury and Death or Recovery of Trees in Prairie Climate. *Ecological Monographs*, Vol. 15, pp. 393-433.

Anderson, R.C.
1970 Prairies in the Prairie State. *Transactions of the Illinois State Academy of Science*, Vol. 63, pp. 214-221.

Anderson, R. C. and M. R. Anderson
1975 The Presettlement Vegetation of Williamson County, Illinois. *Castanea*, Vol. 40, pp. 345-363.

Anderson, R. C. and D. Ugent
1980 Floristic Provinces of Illinois. The Chicago Academy of Science, *Natural History Miscellanea*, No. 210.

Anderson, R. C. and J. White
1970 A Cypress Swamp Outlier in Southern Illinois. *Transactions of the Illinois State Academy of Science*, Vol. 63, pp. 6-13.

Asch, D. L. and N. B. Asch
1975 "Plant Remains from the Zimmerman Site--Grid A: A Quantitative Perspective." *In* The Zimmerman Site: Further Excavations at the Grand Village of the Kaskaskia, by M. K. Brown. *Illinois State Museum Reports of Investigations*, No. 32, pp. 116-120. Springfield.

1977 Chenopod as Cultigen: a Re-evaluation of Some Prehistoric Collections from Eastern North America. *Midcontinental Journal of Archaeology*, Vol. 2, pp. 3-45.

1978a "The Economic Potential of *Iva Annua* and its Prehistoric Importance in the Lower Illinois Valley." *In* The Nature and Status of Ethnobotany, R. I. Ford, (Ed.). *University of Michigan Museum of Anthropology Anthropological Papers*, No. 67, pp. 300-341, Ann Arbor.

1978b "Plant Remains from Frog City: A Havana Site in Southern Illinois." *In* Final Report on Archaeological Investigations at Frog City and Red Light: Two Middle Woodland Period Sites in Alexander County, Illinois, by L. G. Santeford and N. H. Lopinot. *Southern Illinois University Center for Archaeological Investigations Research Paper*, No 6.

1981 "Archaeobotany of Newbridge, Carlin, and Weitzer Sites-The White Hall Components." *In* Faunal Exploitation and Resource Selection: Early Late Woodland Subsistence in the Lower Illinois Valley, by B. W. Styles. *Northwestern University Archaeological Program, Scientific Papers* No. 3, Appendix B.

1982 A Chronology for the Development of Prehistoric in West-central Illinois. Paper Presented at the Forty-seventh Annual meeting of the Society for American Archaeology.

Asch, N. B., R. I. Ford and D. L. Asch
1972 Paleoethnobotany of the Koster Site: The Archaic Horizons. *Illinois State Museum Reports of Investigations*, No. 24, *Illinois Valley Archaeological Program Research Papers*, Vol. 6, Springfield.

Ashby, W. C.
1968 Forest Types of Lusk Creek in Pope County, Illinois. *Transactions of the Illinois State Academy of Science*, Vol. 61, pp. 348-355.

Bell, D. T.
1974 Tree Stratum Composition and Distribution in the Streamside Forest. *American Midland Naturalist*, Vol. 92, pp. 35-46.

Bernabo, J. C.
1981 Quantitative Estimates of Temperature Changes over the Last 2700 Years in Michigan Based on Pollen Data. *Quaternary Research*, Vol. 15, pp. 143-159.

Black, M. J.
1963 The Distribution and Archaeological Significance of the Marshelder, *Iva Annua* L. *Papers of the Michigan Academy of Science, Arts, and Letters*, Vol. 48, pp. 541-547.

1980 Algonquin Ethnobotany: An Interpretation of Aboriginal Adaptation in Southwestern Quebec. *National Museum of Man Canadian Ethnology Service Paper*, No. 65, Ottawa.

Blair, T. A. and R. C. Fite
1965 *Weather Elements: a Text in Elementary Meteorology*. Englewood Cliffs, N. J., Prentice-Hall.

Blake, L. W.
1981 Early Acceptance of Watermelon by Indians of the United States. *Journal of Ethnobiology* Vol.1, pp. 193-199.

Blake, L. W. and Hugh C. Cutler
1975 "Food Plant Remains from the Zimmerman Site." *In* The Zimmerman Site: Further Excavations at the Grand Village of Kaskaskia, by M. K. Brown. *Illinois State Museum Reports of Investigations*, No. 32. Springfield.

Bliss, A.
1978 *Weeds: a Guide For Dyers and Herbalists*. Boulder, Colorado, Juniper House.

Boggess, W. R.
1964 Trelease Woods, Champaign County, Illinois: Woody Vegetation and Stand Composition. *Transactions of the Illinois State Academy of Science*, Vol. 57, pp. 261-271.

Boggess, W.R. and L.W. Bailey

1964 Brownfield Woods, Illinois: Woody Vegetation and Changes Since 1925. *American Midland Naturalist*, Vol. 71, pp. 392-401.

Boggess, W. R. and J. W. Geis

1966 The Funk Forest Natural Area, McLean County, Illinois: Woody Vegetation and Ecological Trends. *Transactions of the Illinois State Academy of Science*, Vol. 59, pp. 123-133.

1967 Composition of An upland, Streamside Forest in Piatt County, Illinois. *The American Midland Naturalist*, Vol. 78, pp. 89-97.

Bolyard, J. L.

1980 *Medicinal Plants and Home Remedies of Appalachia.* Springfield, Ill., Charles C Thomas.

Borchert, J. R.

1950 The Climate of the Central North American Grassland. *Annals of the Association of American Geographers*, Vol. 40, pp. 1-39.

Bourliere, F. and M. Hadley

1970 "Combination of Qualitative and Quantitative Approaches." *In* Analysis of Temperate Forest Ecosystems, D. Reichle (Ed.), New York, Springer-Verlag, pp. 1-6.

Bourdo, E. A

1956 A Review of the General Land Office Survey and of Its Use in Quantitative Studies of Former Forests. *Ecology*, Vol. 37, pp. 754-68.

Braun, E. L.

1947 Development of the Deciduous Forests of Eastern North America. *Ecological Monographs*, Vol. 17, pp. 211-219.

1950 *Deciduous Forests of Eastern North America.* Philadelphia, The Blakiston Company.

Brooklyn Botanic Garden

1964 Dye Plants and Dyeing-a Handbook. *Brooklyn Botanic Garden Handbook* No. 46, *Plants & Gardens*, Vol. 20, No. 3.

1973 Natural Plant Dyeing. *Brooklyn Botanic Garden Handbook* No. 72, *Plants and Gardens*, Vol. 29, No. 2.

Buckman, H. O. and N. C. Brady

1969 *The Nature and Property of Soil.* New York, The MacMillan Company.

Caldwell, J. R.

1958 Trend and Tradition in the Prehistory of the Eastern United States. *American Anthropological Association Memoir*, No. 88.

Calef, R. T.

1953 Flora and Ecological Analysis of the Vegetation of the Funk Forest Natural Area, McLean County, Illinois. *Transactions of the Illinois State Academy of Science*, Vol. 46, pp. 41-55.

Changnon, S. A., S. Hilberg and S. T. Sonka

1982 Crops and Climate: The Variation and Prediction of Yields. *Illinois State Water Survey*, pamphlet.

Christenson, A. L.

1981 "Change in the Human Niche in Response To Population Growth." *In* Modeling Change in Prehistoric Subsistence Economies, T. K. Earle and A. L. Christenson (Eds.). New York, Academic Press, pp. 31-72.

Conard, N., D. L. Asch, N. B. Asch, D. Elmore, H. Gove, M. Rubin, J. Brown, M. Wiant, K. Farnsworth, T. Cook

1984 Accelerator Radiocarbon Dating of Evidence for Prehistoric Horticulture in Illinois. *Nature* Vol. 308, pp. 443-446.

Cottam, G. and J. T. Curtis

1949 A Method For Making Rapid Surveys of Woodlands by Means of Pairs of Randomly Selected Trees. *Ecology*, Vol. 30, pp. 101-104.

1955 Correction For Various Exclusion Angles in the Random Pairs Method. *Ecology*, Vol. 36, p. 767.

1956 The Use of Distance Measures in Phytosociological Sampling. *Ecology*, Vol. 37, pp. 451-460.

Cottam, G., J. T. Curtis and W. B. Hale

1953 Some Sampling Characteristics of a Population of Randomly Dispersed Individuals. *Ecology*, Vol. 34, pp. 741-747.

Cowan, C. W.

1978 "The Prehistoric Use of Maygrass in Eastern North America: Cultural and Phytogeographical Implications." *In* The Nature and Status of Ethnobotany, R. I. Ford (Ed.). *University of Michigan, Museum of Anthropology, Anthropological Papers*, No. 67.

Crites, G. D. and R. D. Terry

1984 Nutritive Value of Maygrass (*Phalaris caroliniana* Walt.) *Economic Botany*, Vol. 38, pp. 114-120.

Crites, R. W. and J. E. Ebinger

1969 Vegetation Survey of Floodplain Forests in East-central Illinois. *Transactions of the Illinois State Academy of Science*, Vol. 62, pp. 316-330.

Croom, E. M.,Jr.

1983 Documenting and Evaluating Herbal Remedies. *Economic Botany*, Vol. 37, pp. 13-27.

Curtis, J. T.

1959 *The Vegetation of Wisconsin: An Ordination of Plant Communities.* Madison, The University of Wisconsin Press.

Cutler, H. C. and L. W. Blake

1976 Plants from Archaeological Sites East of the Rockies. *University of Missouri, American Archaeology Division, American Archaeology Reports*, No. 1. Columbia.

Cutler, H. C. and T. W. Whitaker

1961 History and Distribution of the Cultivated Cucurbits in the Americas. *American Antiquity* Vol. 26, pp. 469-485.

Daubenmire, R. F.

1959 *Plants and Environment: a Textbook of Plant Autecology.* New York, John Wiley & Sons, Inc.

Denmark, W. L.
1974 "The Climate of Illinois." *In* Climates of the States, Vol. I: Eastern States. NOAA, U. S. Department of Commerce, Water Information Center.

Densmore, F.
1928 Uses of Plants By the Chippewa Indians. *Bureau of American Ethnology Annual Report*, No. 44: 275-397.

1929 Chippewa Customs. *Bureau of American Ethnology, Bulletin*, No. 86.

Doebley, J. F.
1984 "Seeds" of Wild Grasses: A Major Food of Southwestern Indians. *Economic Botany*, Vol 38, pp. 52-64.

Earle, T. K.
1981 "A Model of Subsistence Change." *In* Modeling Change in Prehistoric Subsistence Economies, T. K. Earle and A. L. Christenson (Eds.). New York, Academic Press, pp. 1-29.

Ebinger, J. E.
1968 Woody Vegetation Survey of Sargents Woods, Coles County, Illinois. *Transactions of the Illinois State Academy of Science*, Vol. 60, pp. 16-21.

1981 Vegetation of Glacial Drift Hill Prairies in East-central Illinois. *Castanea*, Vol. 46, pp. 115-121.

Ebinger, J. and L. Lehnen
1981 Naturalized Autumn Olive in Illinois. *Transactions of the Illinois State Academy of Science*, Vol. 74, pp. 83-86.

Ebinger, J. E. and H. M. Parker
1969 Vegetation Survey of an Oak-hickory Maple Forest in Clark County, Illinois. *Transactions of the Illinois State Academy of Science*, Vol. 69, pp. 379-387.

Ehrenreich, J. H.
1960 Usable Forage Under Pine Stands? *U.S. Forest Service Central States Forest Experiment Station Note*, No. 142.

Epstein, D. B.
1981 *Plants Used in Pipe Smoking by the Indians of the United States of America East of the Rocky Mountains* Unpublished M.A. thesis, Washington University, St. Louis.

Esau, K.
1960 *Anatomy of Seed Plants*. New York, John Wiley & Sons, Inc.

Evans, D. K.
1979 Floristics of the Middle Mississippi River Sand and Mud Flats. *Castanea*, Vol. 44, pp. 8-24.

Evers, R.
1955 Hill Prairies of Illinois. *Illinois State Natural History Survey Bulletin*, Vol. 26, pp. 367-446.

Fehrenbacher, J. B., B. W. Ray, and J. D. Alexander
1968 "Illinois Soils and Factors in Their Development." *In* The Quaternary of Illinois, R. E. Bergstrom (Ed.). *University of Illinois College of Agriculture Special Publication*, No. 14, pp. 165-175, Urbana.

Fehrenbacher, J. B., G. O. Walker and H. L. Wascher
1967 Soils of Illinois. *University of Illinois Agricultural Experiment Station Bulletin*, No.725.

Fell, E. W. and G. B. Fell
1956 The Gravel-hill Prairies of Rock River Valley in Illinois. *Transactions of the Illinois State Academy of Science*, Vol. 49, pp. 47-62.

Fernald, M. L.
1950 *Gray's Manual of Botany, Eighth Edition*. New York, Van Nostrand Reinhold Company.

Fernald, M. L., A. C. Kinsey, and R. C. Rollins
1958 *Edible Wild Plants of Eastern North America*, Rev. Ed. New York, Harper and Row.

Ford, R. I.
1979 "Paleoethnobotany in American Archaeology." *In* Advances in Archaeological Method and Theory, Vol. 2, M. Schiffer (Ed.). New York, Academic Press.

1981 Gardening and Farming Before A.D. 1000: Patterns of Prehistoric Cultivation North of Mexico. *Journal of Ethnobiology*, Vol. 1, pp. 6-27.

Fowells, H. A.
1965 Silvics of Forest Trees of the United States. *United States Department of Agriculture Forest Service Agriculture Handbook*, No. 271, Washington D. C.

Gates, F. C.
1912 The Vegetation of the Beach Area in Northeastern Illinois and Southeastern Wisconsin. *Illinois Laboratory of Natural History Bulletin*, Vol. 9, pp. 255-372.

Geis, J. W. and W. R. Boggess
1970 Soil-vegetation Relationships in a Prairie Grove Remnant. *Bulletin of the Torrey Botanical Club*, Vol. 97, pp. 196-203.

Gilmore, M. R.
1919 Uses of Plants By the Indians of the Missouri River Region. *In* Smithsonian Institution Bureau of American Ethnology, 33rd Annual Report 1911-1912, pp. 43-154, Washington D. C.

1931 Vegetal Remains of the Ozark Bluff Dwellers Culture. *Papers of the Michigan Academy of Science* Vol.14, pp. 83-103.

1933 Some Chippewa Uses of Plants. *Papers of the Michigan Academy of Science, Arts, and Letters*, Vol. 17, pp. 119-43.

Gleason, H. A.
1910 The Vegetation of the Inland Sand Deposits of Illinois. *Illinois State Laboratory of Natural History Bulletin*, Vol. 9, pp. 23-174.

1913 The Relation of Forest Distribution and Prairie Fires in the Middlewest. *Torreya*, Vol. 13, pp. 173-181.

Glenn-Lewin, D. C. and A. M. Crist
1981 "The Fine Structure of a Prairie Pothole and Pothole Border." *In* The Prairie Peninsula-in the

"Shadow" of Transeau; Proceedings of the Sixth North American Prairie Conference, R. L. Stuckey and K. J. Reese (Eds.), *Ohio Biological Survey Biology Notes*, No. 15.

Griffin, C. D.
1951 Pollen Analysis of a Peat Deposit in Livingston County, Illinois. *Butler University Botanical Studies*, Vol. 10, pp. 90-99.

Griffin, J. B.
1960 Climatic Change: A Contributing Cause of the Growth and Decline of Northern Hopewellian Culture. *The Wisconsin Archeologist*, Vol. 41, pp. 21-33.

Grime, W. E.
1979 *Ethno-botany of the Black Americans*. Algonac, Michigan, Reference Publications, Inc.

Gruger, E.
1972 Pollen and Seed Studies of Wisconsinan Vegetation in Illinois, U.S.A. *Geological Society of America Bulletin*, Vol. 83, pp. 2715-2734.

Hargrave, M. L. and B. M. Butler
1981 "Plant Remains." *In* Archaeological Testing of the Bridges Site, Mr-11, Marion County, Illinois, M. L. Hargrave and B. M. Butler. *Southern Illinois University Center for Archaeological Investigations Research Paper*, No. 30.

Harn, A. D. and A. C. Koelling
1974 The Indigenous Drug Plants of Fulton County, Illinois. *Transactions of the Illinois State Academy of Science*, Vol. 67, pp. 259-284.

Harshberger, J. W.
1896 The Purpose of Ethnobotany. *American Antiquarian*, Vol. 17, pp. 73-81.

Heineke, T. E.
1978 The Vascular Flora of Cedar Lake Reservoir, Jackson County, Illinois. *Transactions of the Illinois State Academy of Science*, Vol. 71, pp. 126-155.

Hosner, J. F. and L. S. Minckler
1963 Bottomland Hardwood Forests of Southern Illinois--Regeneration and Succession. *Ecology*, Vol. 44, pp. 29-41.

Hudnut, R.
1952 A Mesophytic Forest on the Upland Prairie. *Transactions of the Illinois State Academy of Science*, Vol. 45, pp. 52-47.

Hughes, J. T. and J. E. Ebinger
1973 Woody Vegetation Survey of Rocky Branch Nature Preserve, Clark County, Illinois. *Transactions of the Illinois State Academy of Science*, Vol. 66, pp. 44-54.

Jackson, M. T.
1977 Composition of the Presettlement Forests of the Unglaciated Jo Davies Hills of Northwestern Illinois, USA. *Transactions of the Missouri Academy of Science*, Vol. 10/11, p. 301.

Jackson, M. T. and R. O. Petty
1971 An Assessment of Various Synthetic Indices in a Transitional Old-growth Forest. *American Midland Naturalist*, Vol. 86, pp. 13-27.

Jackson, R. C.
1960 A Revision of the Genus *Iva* L. *University of Kansas Science Bulletin*, Vol. 41, pp. 793-876.

Johannessen, S.
1981 "Plant Remains from the Lohmann Site (11-S-49)." *In* Final Report on FAI 270 and Illinois Route 460 Related Excavations at the Lohmann Site 11-S-49, by D. Esarey and T. W. Good. *FAI 270 Archaeological Mitigation Project Report* No. 39, pp. 186-196.

1982 "Paleoethnobotany" *In* Annual Report of 1981 Investigations by the University of Illinois At Urbana-Champaign FAI-270 Archaeological Mitigation Project, pp. 36-38.

1983a "Plant Remains from the Missouri Pacific #2 Site." *In* The Missouri Pacific #2 Site (11-2-46), by Dale I. McElrath and A. C. Fortier. *American Bottom Archaeology FAI-270 Site Reports*, Vol. 3, pp. 191-207.

1983b "Floral Remains from the Early Woodland Florence Phase." *In* The Florence Street Site (11-S-458), by T. E. Emerson, G. R. Milner and D. K. Jackson. *American Bottom Archaeology FAI-270 Site Reports*, Vol. 2, pp. 133-146.

1983c "Paleoethnobotany" *In* Annual Report of 1982 Investigations by the University of Illinois at Urbana-Champaign FAI-270 Archaeological Mitigation Project, pp. 24-25.

1984 "Paleoethnobotany." *In* American Bottom Archaeology: A Summary of the FAI-270 Project Contribution to the Culture History of the Mississippi River Valley, C. J. Bareis and J. W. Porter (Eds.). University of Illinois Press, Urbana.

Johannessen, S. and L. Whalley
1981 "Annual Report of Paleoethnobotanical Investigations." *In* Annual Report of 1981 Investigations by the University of Illinois at Urbana-Champaign FAI-270 Archaeological Mitigation Project, pp. 154-157.

Johnson, F. L.
1974 Tree Size Is Clue To Past History of Woodlands. *Illinois Research*, Vol. 16, No. 3, p. 18.

Jones, A. G. and D. T. Bell
1974 Vascular Plants of the Sangamon River Basin: Annotated Checklist and Ecological Summary. *University of Illinois College of Agriculture Agricultural Experiment Station Bulletin*, No. 746, Urbana.

Jones, D. M.
1966 Variability of Evapotranspiration in Illinois. *Illinois State Water Survey, Circular*, No. 89.

Jones, G. N.
1950 Flora of Illinois. *The American Midland Naturalist Monograph*, No. 5.

Jones, G. N. and G. D. Fuller
1955 Vascular Plants of Illinois. *Illinois State Museum Scientific Series*, Vol. 6, Springfield.

Kay, M., F. B. King and C. K. Robinson
1980 Cucurbits from Phillips Spring: New Evidence and Interpretations. *American Antiquity*, Vol. 45, pp. 806-822.

King, F. B.
1978 Additional Cautions on the Use of the GLO Survey Records in Vegetational Reconstructions in the Midwest. *American Antiquity*, Vol. 43, pp. 99-103.

1980 "Plant Remains from Labras Lake." *In* Investigations at the Labras Lake Site: Volume I-Archaeology, by J. L. Phillips, R. L. Hall, and R. W. Yerkes, pp. 325-337. *University of Illinois at Chicago Circle Department of Anthropology Reports of Investigations*, No. 1.

1981 "Analysis of Plant Remains from the 1980 Excavations at Modoc." *In* Modoc Rock Shelter Archaeological Project, Randolph County, Illinois, 1980-81, by B. W. Styles, M. L. Fowler, S. R. Ahler, F. B. King, and T. R. Styles. Completion Report to the Department of the Interior, Heritage Conservation and Recreation Service, and the Illinois Department of Conservation.

1983 *Archaeobotanical Remains from the Rench Site*. Paper presented at the Forty-eighth Annual Meeting of the Society for American Archaeology.

King, F. B. and R. W. Graham
1981 Effects of Ecological and Paleoecological Patterns on Subsistence and Paleoenvironmental Reconstructions. *American Antiquity*, Vol. 48, No. 1, pp. 128-142.

King, F. B. and J. B. Johnson
1977 Presettlement Forest Composition of the Central Sangamon Basin. *Transactions of the Illinois State Academy of Science*, Vol. 70, pp. 153-163.

King, F. B. and D. Roper
1976 Floral Remains from Two Middle to Early Late Woodland Sites in Central Illinois and Their Implications. *The Wisconsin Archeologist*, Vol. 57, pp. 142-151.

King, J. E.
1981 Late Quaternary Vegetational History of Illinois. *Ecological Monographs*, Vol. 51, pp. 43-62.

King, J. E., J. A. Lineback, and D. L. Gross
1976 Palynology and Sedimentology of Holocene Deposits in Southern Lake Michigan. *Illinois State Geological Survey Circular*, No. 496.

Kingsbury, J. M.
1964 *Poisonous Plants of the United States and Canada*. Prentice-Hall, Englewood Cliffs, N.J.

Kurz, D. R.
1981 "Flora of Limestone Glades in Illinois." *In* The Prairie Peninsula--in the "Shadow" of Transeau; Proceedings of the Sixth North American Prairie Conference, R. L. Stuckey and K. J. Reese (Eds.), *Ohio Biological Survey Biology Notes*, No. 15.

Leighton, M. M., G. E. Ekblaw, and C. L. Horberg
1948 Physiographic Divisions of Illinois. *Journal of Geology*, Vol. 56, pp. 16-33.

Leitner, L. A. and M. T. Jackson
1981 Presettlement Forests of the Unglaciated Portion of Southern Illinois. *American Midland Naturalist*, Vol. 105, pp. 290-304.

Lewis, W. H. and M. Elvin-Lewis
1977 *Medical Botany: Plants Affecting Man's Health*. New York, John Wiley & Sons.

Lindsey, A. A., R. O. Petty, D. K. Sterling, and W. Van Asdall
1961 Vegetation and Environment Along the Wabash and Tippecanoe Rivers. *Ecological Monographs*, Vol. 31, pp. 105-156.

Lineback, J. A.
1981 Quaternary Deposits of Illinois. *Illinois State Geological Survey*, map.

Lopinot, N. H.
1981a "Evaluation of Flotation Samples from the Bridges Site." *In* Archaeological Testing of the Bridges Site, Mr-11, Marion County, Illinois, by M.L. Hargrave and B.M. Butler. *Southern Illinois University Center for Archaeological Investigations Research Paper*, No. 30, pp. 75-79.

1981b "Evaluation of Flotation Samples from the Bridges Site." *In* Archaeological Testing of the Bridges Site, Mr-11, Marion County, Illinois, by M.L. Hargrave and B.M. Butler. *Southern Illinois University Center for Archaeological Investigations Reserach Paper*, No. 30, pp. 68-75.

1982 "Plant Remains." *In* Archaeological Investigations in the Turkey Bluffs Fish and Wildlife Area, Randolph County, Illinois, by J.S. Penny, Jr. *Southern Illinois University Center for Archaeological Investigations Research Paper*, No. 35.

1983 "Analysis of Flotation Sample Materials from the Late Archaic Horizon." *In* The 1982 Excavations at the Cahokia Interpretive Center Tract, St. Clair County, Illinois, by M.S. Nassaney, N.H. Lopinot, B.M. Butler, and R.W. Jefferies. *Southern Illinois University Center for Archaeological Investigations Research Paper*, No. 37.

Lust, J.
1974 *The Herb Book*. New York, Bantam Books, Inc.

McClain, W. E. and J. E. Ebinger
1968 Woody Vegetation of Baber Woods, Edgar County, Illinois. *American Midland Naturalist*, Vol. 79, pp. 419-428.

McMillan, R. B. and W. E. Klippel
1981 Post-glacial Environmental Change and Hunting-gathering Societies of the Southern Prairie Peninsula. *Journal of Archaeological Science*, Vol. 8, pp. 215-245.

Madany, M. H.
1981 "A Floristic Survey of Savannas in Illinois." *In* The Prairie Peninsula-in the "Shadow" of Transeau; Proceedings of the Sixth North American Prairie Conference, R. L. Stuckey and K. J. Reese (Eds.), *Ohio Biological Survey Biology Notes*, No. 15.

Mohlenbrock, R. H.
1959 A Floristic Study of a Southern Illinois Swampy Area. *Ohio Journal of Science*, Vol. 59, pp. 89–100.

1967– *The Illustrated Flora of Illinois*. Carbondale, Southern Illinois University Press. (Series)

1967 A Floristics Study of Lake Murphysboro State Park, Illinois. *Transactions of the Illinois State Academy of Science*, Vol. 60, pp. 409–421.

1975 *Guide to the Vascular Flora of Illinois*. Carbondale, Southern Illinois University Press.

Mohlenbrock, R. H., G. E. Dillard, and T. S. Abney
1961 A Survey of Southern Illinois Aquatic Vascular Plants. *Ohio Journal of Science*, Vol. 61, pp. 262–273.

Mohlenbrock, R. H. and D. M. Ladd
1978 *Distribution of Illinois Vascular Plants*. Carbondale, Southern Illinois University Press.

Mohlenbrock, R. H. and J. W. Voigt
1965 An Annotated Checklist of Vascular Plants of the Southern Illinois University Pine Hills Field Station and Environs. *Transactions of the Illinois State Academy of Science*, Vol. 58, pp. 268–301.

Moran, R. C.
1976 "Presettlement Vegetation of Lake County, Illinois." *In* Proceedings of the Fifth Midwest Prairie Conference, D. C. Glenn-Lewin and R. Q. Landers, Jr. (Eds.). Ames, Iowa, Iowa State University.

1981 "Prairie Fens in Northeastern Illinois: Floristic Composition and Disturbance." *In* The Prairie Peninsula-in the "Shadow" of Transeau; Proceedings of the Sixth North American Prairie Conference, R. L. Stuckey and K. J. Reese (Eds.), *Ohio Biological Survey Biology Notes*, No. 15.

Munson, P. J., P. W. Parmalee, and R. A. Yarnell
1971 Subsistence Ecology of Scovill, A Terminal Middle Woodland Village. *American Antiquity*, Vol. 36, pp. 410–431.

Nyboer, R. W. and J. E. Ebinger
1976 Woody Vegetation Survey of a Terrace Forest in East-central Illinois. *Castanea*, Vol. 41, pp. 348–356.

Oosting, H. J.
1956 *The Study of Plant Communities: An Introduction to Plant Ecology*. San Francisco, W. H. Freeman and Company.

Ozment, J. E.
1967 The Vegetation of Limestone Ledges of Southern Illinois. *Transactions of the Illinois State Academy of Science*, Vol. 60, pp. 135–173.

Page, J. L.
1949 Climate of Illinois. *University of Illinois College of Agriculture Agriculture Experiment Station Bulletin*, No. 532.

Parker, A. C.
1910 Iroquois Uses of Maize and Other Food Plants. *New York State Museum Bulletin* 144.

Payne, W. W., and V. H. Jones
1962 The Taxonomic Status and Archeological Significance of a Giant Ragweed from Prehistoric Bluff Shelters in the Ozark Plateau. *Papers of the Michigan Academy of Science, Arts, and Letters*, Vol. 47, pp. 147–163.

Peterson, G. M.
1976 Pollen Spectra from Surface Sediments of Lakes and Ponds in Kentucky, Illinois and Missouri. *American Midland Naturalist*, Vol. 100, pp. 333–340.

Piskin, K. and R. E. Bergstrom
1967 Glacial Drift in Illinois: Thickness and Character. *Illinois State Geological Survey Circular*, No. 416.

Phillippe, P. E. and J. E. Ebinger
1973 Vegetation Survey of Some Lowland Forests Along the Wabash River. *Castanea*, Vol. 38, pp. 339–349.

1977 Vascular Plants in Walnut Point State Park, Douglas County, Illinois. *Transactions of the Illinois State Academy of Science*, Vol. 69, pp. 437–445.

Phipps, R. and J. Speer
1958 A Hillside Marsh in East-central Illinois. *Transactions of the Illinois State Academy of Science*, Vol. 51, pp. 37–42.

Rapport, D. J. and J. E. Turner
1977 Economic Models in Ecology. *Science*, Vol. 195, pp. 367–373.

Reagan, A. B.
1928 Plants Used by The Bois Fort Chippewa. *Wisconsin Archeologist*, n.s. Vol. 7, pp. 230–248.

Reed, C. A.
1977 "The Origins of Agriculture: Prologue." *In* The Origins of Agriculture, C. A. Reed (Ed.). The Hague, Mouton Publishers, Inc.

Robertson, P. A., G. T. Weaver, and J. A. Cavanaugh
1978 Vegetation and Tree Species Patterns Near the Northern Terminus of the Southern Floodplain Forest. *Ecological Monographs*, Vol. 48, pp. 249–267.

Rodgers, C. S. and R. C. Anderson
1979 Presettlement Vegetation of Two Prairie Peninsula Counties. *Botanical Gazette*, Vol. 140, pp. 232–240.

Root, T. W., J. W. Geis and W. R. Boggess
1971 Woody Vegetation of Hart Memorial Woods, Champaign County, Illinois. *Transactions of the Illinois State Academy of Science*, Vol. 64, pp. 27–37.

Sampson, H. C.
1921 An Ecological Survey of the Prairie Vegetation of Illinois. *Illinois Natural History Survey Bulletin*, No. 13, pp. 523–576.

Schwegman, J. E., G. B. Fell, M. Hutchison, G. Paulson, W. M. Shepherd, and J. White
1973 *Comprehensive Plan for the Illinois Nature Preserves System, Part 2: The Natural Divisions of Illinois*. Rockford, Illinois Nature Preserves Commission.

Sheldon, E.
1980 *Prunus persica*. Paper presented at the Twenty-first Annual Meeting of the Society for Economic Botany, June 15-18, 1980. Indiana University, Bloomington, Indiana.

Shelford, V. E.
1974 *The Ecology of North America*. Urbana, University of Illinois Press.

Sherff, E. E.
1913 Vegetation of Skokie Marsh. *State Laboratory of Natural History Bulletin*, Vol. 9, pp. 575-614.

Sheviak, C. J. and A. Haney
1973 Ecological Interpretations of the Vegetation Patterns of Volo Bog, Lake County, Illinois. *Transactions of the Illinois State Academy of Science*, Vol. 66, pp. 99-112.

Smith, B. D.
1975 Middle Mississippi Exploitation of Animal Populations. *University Michigan Museum of Anthropology, Anthropological Papers*, No. 57.

Smith, H. H.
1923 Ethnobotany of the Menomini Indians. *Bulletin of the Milwaukee Public Museum*, Vol. 4, pp. 1-174.

1928 Ethnobotany of the Meskwaki Indians. *Bulletin of the Milwaukee Public Museum*, Vol. 4, pp. 175-326.

1932 Ethnobotany of the Ojibwe Indians. *Bulletin of the Milwaukee Public Museum*, Vol. 4, pp. 327-525.

1933 Ethnobotany of the Forest Potawatomi Indians. *Bulletin of the Milwaukee Public Museum*, Vol. 7, pp. 1-230.

Smith, P. W.
1961 The Amphibians and Reptiles of Illinois. *Illinois Natural History Survey Bulletin*, Vol 28, No. 1.

Solomon, J. C.
1979 An Annotated List of Vascular Plants from Knox County, Illinois. *Transactions of the Illinois State Academy of Science*, Vol. 72, pp. 9-29.

Stewart, L. O.
1935 *Public Land Surveys*. Ames, Iowa, Collegiate Press, Inc.

Stoltman, J. B.
1978 Temporal Models in Prehistory: An Example from Eastern North America. *Current Anthropology*, Vol. 19, pp. 703-729.

Struever, S.
1968a Flotation Techniques for the Recovery of Small-scale Archaeological Remains. *American Antiquity*, Vol. 33, pp. 353- 362.

1968b "Woodland Subsistence-Settlement Systems in the Lower Illinois Valley." *In* New Perspectives in Archeology, S.R. Binford and L.R. Binford (Eds.), pp. 285-312. Chicago, Aldine.

Styles, B. W.
1981 Faunal Exploitation and Resource Selection: Early Late Woodland Subsistence in the Lower Illinois Valley. *Northwestern University Archaeological Program Scientific Papers*, No. 3.

Styles, B. W., S. Ahler, and M. L. Fowler
1983 "Modoc Rock Shelter Revisited." *In* Archaic Hunters and Gatherers in the American Midwest. J. Brown and J. Phillips (Eds.). New York, Academic Press, pp. 261-297.

Swanton, J. R.
1946 The Indians of the Southeastern United States. *Smithsonian Institution Bureau of American Ethnology Bulletin*, No. 137.

Tehon, L. R.
1951 The Drug Plants of Illinois. *Illinois State Natural History Survey Circular*, No. 44.

Telford, C. J.
1926 Third Report of a Forest Survey of Illinois. *Illinois State Natural History Bulletin*, No. 16.

Transeau, E. N.
1935 The Prairie Peninsula. *Ecology*, Vol. 16, pp. 423-437.

Trowbridge, C. C.
1938 Meearmeear Traditions. Edited by V. Kinietz. *University of Michigan, Museum of Anthropology, Occasional Contribution*, No. 7.

Turner, L. M.
1934 Grassland in the Floodplain of Illinois Rivers. *American Midland Naturalist*, Vol. 15, pp. 770-780.

1936 Ecological Studies in the Lower Illinois River Valley. *Botanical Gazette*, Vol. 97, pp. 689-727.

Tyler, V. E.
1982 *The Honest Herbal: A Sensible Guide to Herbs and Related Medicines*. Philadelphia, George F. Stickley Company.

Vestal, A. G.
1931 A Preliminary Vegetation Map of Illinois. *Transactions of the Illinois State Academy of Science*, Vol. 23, pp. 204-217.

1936 Barrens Vegetation in Illinois. *Transactions of the Illinois State Academy of Science*, Vol. 29, pp. 79-80.

Vogel, R. L. and J. E. Ebinger
1979 Frequency of Aquatic Macrophytes in East-central Illinois. *Transactions of the Illinois State Academy of Science*, Vol. 72, pp. 37-41.

Vogel, V. J.
1970 *American Indian Medicine*. Norman, University of Oklahoma Press.

Voigt, J. W. and R. H. Mohlenbrock
1964 *Plant Communities of Southern Illinois*. Carbondale, Illinois University Press.

[1979] *Prairie Plants of Illinois*. Illinois Department of Conservation.

Voss, J.
1937 Comparative Study of Bogs on Cary and Tazewell Drift in Illinois. *Ecology*, Vol. 18, pp. 119-135.

Watson, P.J.
1976 In Pursuit of Prehistoric Subsistence: A Comparative Account of Some Contemporary Flotation Techniques. *Midcontinental Journal of Archaeology*, Vol. 1, pp. 77-100.

Watt, B. K. and A. L. Merrill
1963 Composition of Foods. *United States Department of Agriculture Handbook*, No. 8. Washington D. C.

Waugh, F. W.
1916 Iroquois Foods and Food Preparation. *Canada Department of Mines Geological Survey Anthropological Series*, No. 12.

Weaver, G. T. and W. C. Ashby
1971 Composition and Structure of an Old-growth Forest Remnant in Unglaciated Southwestern Illinois. *Transactions of the Illinois State Academy of Science*, Vol. 86, pp. 46-56.

Weik, K. L. and G. Baker
1975 A Preliminary Survey of Vascular Plants of the Horseshoe Lake Area, Knox County, Illinois. *Transactions of the Illinois State Academy of Science*, Vol. 68, pp. 381-388.

Wells, P. V. and G. E. Morley
1964 Composition of Baldwin Woods: An Oak-hickory Forest in Eastern Kansas. *Transactions of the Kansas Academy of Science*, Vol. 67, pp. 65-69.

Wendland, W. M. and R. A. Bryson
1974 Dating Climatic Episodes of the Holocene. *Quaternary Research*, Vol. 4, pp. 9-24.

Whalley, L. A.
1983 "Plant Remains from the Turner Site." *In* The Turner and DeMange Sites (11-S-50) (11-S-447), by G.R. Milner and J.A. Williams. *American Bottom Archaeology FAI-270 Site Reports*, Vol. 4.

White, J. and M. H. Madany
1978 "Classification of Natural Communities in Illinois." *In* Illinois Natural Areas Inventory Technical Report, Volume I: Survey Methods and Results. Urbana,Illinois Natural Areas Inventory, pp. 311-405.

1981 "Classification of Prairie Communities in Illinois." *In* The Prairie Peninsula-in the "Shadow" of Transeau; Proceedings of the Sixth North American Prairie Conference, R. L. Stuckey and K. J. Reese (Eds.). *Ohio Biological Survey Biology Notes*, No. 15, pp. 169-171.

Whitford, A. C.
1941 Textile Fibers used in Eastern Aboriginal North America. *Anthropological Papers of the American Museum of Natural History*, Vol. 38, Part 1, pp. 5-22.

Whittaker, R. H.
1975 *Communities and Ecosystems*, 2nd New York, McMillan Publishing Co., Inc.

Willman, H. B., E. Atherton, T. C. Buschbach, C. Collinson, J. C. Frye, M. E. Hopkins, J. A. Lineback, J. A. Simon
1975 Handbook of Illinois Stratigraphy. *Illinois State Geological Survey Bulletin*, No. 95.

Willman, H. B. and J. C. Frye
1970 Pleistocene Stratigraphy of Illinois. *Illinois State Geological Survey Bulletin*, No. 94.

Wilson, H. D.
1981 Domesticated *Chenopodium* of the Ozark Bluff Dwellers. *Economic Botany*, Vol. 35, pp. 233-239.

Winterringer, G. S. and R. A. Evers
1960 New Records For Illinois Vascular Plants. *Illinois State Museum Scientific Paper Series*, Vol. 11. Springfield.

Winterringer, G. S. and A. C. Lopinot
1966 Aquatic Plants of Illinois. *Illinois State Museum Popular Science Series*, Vol. 6. Springfield.

Winterringer, G. S. and A. G. Vestal
1956 Rock-ledge Vegetation in Southern Illinois. *Ecological Monographs*, Vol. 26, pp. 105-130.

Wood, W. R.
1976 Vegetational Reconstructions and Climatic Episodes. *American Antiquity*, Vol. 41, pp. 206-07.

Wright, H. E., Jr.
1968 "History of the Prairie Peninsula." *In* The Quaternary of Illinois, R. E. Bergstrom (Ed.). *University of Illinois College of Agriculture Special Publication*, No. 14, pp. 78-88. Urbana.

Wunderlin, R. P.
1966 The Vascular Flora of the Mississippi Palisades State Park, Carroll County, Illinois. *Transactions of the Illinois State Academy of Science*, Vol. 59, pp. 134-148.

Yanovsky, E.
1936 Food Plant of the North American Indians. *United States Department of Agriculture Miscellaneous Publications*, No. 237.

Yarnell, R. A.
1964 Aboriginal Relationships Between Culture and Plant Life in the Upper Great Lakes Region. *University of Michigan Museum of Anthropology, Anthropological Papers*, No. 23.

1972 *Iva annua* var. *macrocarpa: Extinct American Cultigen? American Anthropologist*, Vol. 74, pp. 335-341.

1977 "Plant Husbandry North of Mexico." *In* Origins of Agriculture, C. A. Reed (Ed.). The Hague, Mouton Publishers.

1978 "Domestication of Sunflower and Sumpweed in Eastern North America." *In* The Nature and Status of Ethnobotany, R. I. Ford (Ed). *University of Michigan Museum of Anthropology, Anthropological Papers*, No. 67, pp. 289-299.

1983 Prehistoric Plant Foods and Husbandry in Eastern North America. Paper presented at the Forty-eighth Annual Meeting of the Society for American Archaeology.

Zawacki, A. A. and G. Hausfater
1969 Early Vegetation of the Lower Illinois Valley. *Illinois State Museum, Reports of Investigations*, No. 17. Springfield.

GLOSSARY OF MEDICAL TERMS

ABORTIFACIENT - a drug that causes abortion.
AGUE - a fit or spell of shaking or shivering (as with a cold).
ALTERATIVE - an agent causing gradual restoration of health.
ANALGESIC - a substance that produces insensitivity to pain without the loss of consciousness.
ANODYNE - an agent that relieves pain or soothes.
ANTHELMINTHIC - an agent that kills or expels intestinal worms.
ANTIPHLOGISTIC - an agent that relieves inflammation.
ANTIPRURITIC - a substance that tends to check or alleviate itching.
ANTIPYRETIC - agent relieving or reducing fever.
ANTISPASMODIC - an agent relieving or preventing spasms.
APERIENT - a laxative.
AROMATIC - a plant, drug, or medicine characterized by a fragrant smell and usually by a warm, pungent taste.
ASTRINGENT - a medicine for checking the discharge of mucous or serum by causing shrinkage of tissue.
CARMINATIVE - an agent that expels gas from the alimentary canal, relieving colic, griping, or flatulence.
CATARRH - inflammation of a mucous membrane, especially of the nose or throat, causing an increased flow of mucous.
CATHARTIC - having the effect of cleansing or purifying.
CONSUMPTION - tuberculosis of the lungs.
COUNTERIRRITANT - anything used to produce a slight irritation in order to relieve a more serious irritation elsewhere.
DEMULCENT - a substance capable of soothing an inflamed or abraded mucous membrane or protecting it from irritation.
DIAPHORETIC - an agent inducing sweating.
DIURETIC - an agent that increases the flow of urine.
DROPSY - edema, swelling due to water retention.
DYSMENORRHEA - painful menstruation.
DYSURIA - painful urination.
ECZEMA - a skin disease characterized by inflammation, itching, and scales.
EMBROCATION - a liniment.
EMETIC - an agent that induces vomiting.
EMMENAGOGUE - an agent that promotes menstrual discharge.
EMOLLIENT - a preparation applied, usually to the skin, to make it soft and supple.
ESCHAROTIC - an agent that produces a scab, especially one formed after a burn.
EXPECTORANT - an agent that facilitates or promotes discharge of mucous from the respiratory tract.
FEBRIFUGE - an agent that mitigates or removes fever.
FLUX - an excessive or abnormal discharge of fluid from the body, usually from the bowels; an agent to cause such a discharge.
HEMORRHAGE - excessive bleeding.
HEMOSTATIC - an agent that arrests bleeding.
HYDRAGOGUE - a cathartic that causes copious water discharge from the bowels.
HYDROPHOBIA - rabies.
LEUKORRHEA - a discharge from the vagina due to inflammation or congestion of the uterine or vaginal mucous membrane.
NERVINE - an agent soothing nervous excitement.
NEURALGIA - severe pain along a nerve.
PARTURIFACIENT - producing or easing labor in childbirth.
PECTORAL - something used in treatment of diseases of the chest and lungs.
PHYSIC - a cathartic, usually for the bowels.
PLEURISY - inflammation of the lining of the chest cavity causing painful breathing.
POULTICE - a soft mass applied to sores or inflamed areas to supply moist warmth, relieve pain or to act as a counterirritant or antiseptic.
PURGATIVE - a cathartic.
QUINSY - tonsillitis, sore throat.
REFRIGERANT - an agent used to reduce fever.
REVIVER - an agent restoring life or consciousness.
RUBEFACIENT - an agent causing redness of the skin.
SEDATIVE - an agent lessening excitement, nervousness, or irritation.
STOMACHIC - a tonic or stimulant for the stomach.
STERNUTATORY - an agent which induces sneezing.
STYPTIC - an agent that tends to check bleeding.
SUDORIFIC - an agent causing or inducing sweating.
TAENIAFUGE - an agent that expels tapeworms.
THRUSH - any of several oral disorders.
VERMIFUGE - serving to destroy or expel worms, especially of the intestines.
VERTIGO - a sensation of dizziness.

INDEX OF SCIENTIFIC NAMES USED IN THE LISTING OF VASCULAR PLANTS

INDEX OF COMMON NAMES USED IN THE LISTING OF VASCULAR PLANTS